C000137877

Damon Cooper
Business Development
Do Not Remove

Energy-Efficient
Building Systems

ABOUT THE AUTHOR

Dr. Lal Jayamaha is the Director of the United PREMAS
Energy Centre in Singapore. He has been involved in many
energy-saving projects, and regularly conducts seminars
and delivers lectures on energy issues in Singapore, Hong
Kong, Malaysia, China, Thailand, and Indonesia.
Dr. Jayamaha earned a Ph.D. degree from the National
University of Singapore, and is a Chartered Mechanical
Engineer (UK) and a Professional Engineer (Singapore).
He is a member of the Institution of Mechanical Engineers
(UK) and the American Society of Heating, Refrigerating
& Air-Conditioning Engineers (ASHRAE).

Energy-Efficient Building Systems

**Green Strategies for
Operation and Maintenance**

Dr. Lal Jayamaha

McGraw-Hill

New York   Chicago   San Francisco   Lisbon   London   Madrid
Mexico City   Milan   New Delhi   San Juan   Seoul
Singapore   Sydney   Toronto

The McGraw-Hill Companies

Library of Congress Cataloging-in-Publication Data

Jayamaha, Lal.
   Energy-efficient building systems : green strategies for operation
and maintenance / Lal Jayamaha.
      p.   cm.
   Includes bibliographical references and index.
   ISBN 0-07-148282-2 (alk. paper)
   1. Buildings—Energy conservation.   I. Title.

   TJ163.5.B84J39   2006
   697—dc22   2006033794

Copyright © 2007 by The McGraw-Hill Companies, Inc. All rights
reserved. Printed in the United States of America. Except as permitted
under the United States Copyright Act of 1976, no part of this publica-
tion may be reproduced or distributed in any form or by any means, or
stored in a data base or retrieval system, without the prior written per-
mission of the publisher.

4 5 6 7 8 9 10 11 12   IBT/IBT   1 9 8 7 6 5 4 3 2 1 0

ISBN-13: 978-0-07-148282-0
ISBN-10: 0-07-148282-2

*The sponsoring editor for this book was Cary Sullivan and the
production supervisor was Richard C. Ruzycka. It was set in Century
Schoolbook by International Typesetting and Composition. The art
director for the cover was Anthony Landi.*

Printed and bound by IBT Global.

This book is printed on acid-free paper.

McGraw-Hill books are available at special quantity discounts to use as
premiums and sales promotions, or for use in corporate training programs. For more information, please write to the Director of Special Sales,
McGraw-Hill Professional, Two Penn Plaza, New York, NY 10121-2298.
Or contact your local bookstore.

Information contained in this work has been obtained by The McGraw-Hill
Companies, Inc. ("McGraw-Hill") from sources believed to be reliable.
However, neither McGraw-Hill nor its authors guarantee the accuracy or
completeness of any information published herein, and neither McGraw-Hill
nor its authors shall be responsible for any errors, omissions, or damages
arising out of use of this information. This work is published with the
understanding that McGraw-Hill and its authors are supplying information but are not attempting to render engineering or other professional
services. If such services are required, the assistance of an appropriate professional should be sought.

This book is dedicated to the memory of my parents; to my wife Kshani, for her love and encouragement; and to sons Michael and Andrew, for their inspiration and support.

Contents

Preface

Energy in the form of electricity, oil, and gas is used in buildings for operating systems such as air-conditioning, heating, ventilation, lighting, and vertical transportation, which are essential for ensuring the safety and comfort of the building's occupants. These systems account for 70 to 80 percent of the total energy consumed in buildings. Energy costs roughly account for about 30 to 40 percent of the total operating cost of a typical building. Therefore, as energy prices soar, more building owners and operators are turning to energy management to trim their overall operating costs. *Energy management* includes improving the *energy efficiency* of building systems and *energy conservation,* involving cutting down on energy wastage, which, based on past experience, is able to save as much as 30 percent of the annual energy cost of buildings.

Further, in most countries, electricity, which is one of the main forms of energy used in buildings, is generated using fossil fuels such as oil, gas, and coal. These fossil fuels are nonrenewable and also, during combustion emit carbon dioxide, which contributes to global warming. Since buildings typically account for more than one-third of the total national energy consumption, government agencies in many countries are also promoting energy management as a means to control energy resources and environmental emissions.

This book on energy efficient building systems is written mainly for two categories of readers. The first category includes professionals who are involved in the operation and maintenance of buildings and are interested in reducing energy costs. The other category of readers are students who are specializing in this subject and professionals working in energy service companies (ESCOs), who want to broaden their knowledge about the subject.

The book is written based on the author's personal experience; having been involved in research and energy saving projects for over 15 years. It contains ten self-contained chapters on the different aspects and systems relating to energy consumption in buildings. The book is intended to serve as a "practical" guide for readers interested in energy savings and contain many actual case studies and examples from the author's past work. Further, the energy saving strategies and measures described in the book have been implemented before and, therefore, are of a practical nature and can be applied by readers based on their individual needs.

Chapter 1 provides an introduction to the subject of energy management and energy audits and illustrates the use of energy audits to identify various energy saving opportunities in buildings. Thereafter, Chapters 2 to 10 describe the different systems and aspects of energy consumption in buildings. Each chapter includes a brief introduction to the subject to help readers who may not be familiar with some of the basic concepts associated with each topic. Most chapters also include a few review questions for students who may use the book as an accompanying text in undergraduate and postgraduate studies. Worked solutions have also been provided to assist them.

Chapter 2 on central air-conditioning systems provides a brief introduction to refrigeration cycles and chiller operations, and includes a number of energy management strategies, which cover, both, design and operational aspects of chiller systems. Chapter 3 provides an overview of boilers and how they operate, followed by potential energy saving measures that can improve the efficiency of boilers through operational strategies and system design. Other opportunities for improving the efficiency of boiler systems, including optimization measures for ancillary equipment are also described in detail. Chapter 4 on pumping systems provides an overview of the pumping systems used in buildings for air-conditioning and heating systems, followed by various energy management strategies for chilled water, hot water, and condenser water pumping systems. An overview of cooling towers and their application in buildings, followed by a number of energy saving strategies to ensure optimum efficiency of air-conditioning systems is described in Chapter 5. Thereafter, various design and operational strategies for reducing energy consumption by improving the energy efficiency of air handling and air distribution systems are described in Chapter 6. Next, various energy saving measures relating to building lighting systems, such as optimizing lighting levels, improving lighting system efficiency, lighting controls, and daylighting, are presented in Chapter 7. Utility companies normally charge for electricity use based on energy consumption, power demand, and power factor, and various strategies to reduce these electricity charges are discussed in Chapter 8. Energy consumed by building systems, such as air-conditioning, heating, and lighting systems, can be reduced by ensuring that their operations are optimized by implementing various control strategies; and Chapter 9 illustrates how such strategies can be implemented using the various functions and features of building automation systems. Finally, Chapter 10 outlines strategies that can be used to reduce the overall energy consumption of buildings by improving the design of building envelopes.

I would like to thank my present and past colleagues without whom I would not have been able to successfully complete so many energy audits and energy saving projects. Special thanks to Adrian Wang, who over the years has been continuously reminding me that I should write a book when I retire (I couldn't wait that long!). I would also like to thank Goh Chwee Guan (York International), Teh Eng Chuan (Philips), Kim Jae Soo (EnE System), Steve Connor (Cleaver Brooks), Phil Stockford (Spirax Sarco), Jasmine Williams (Calmac), Robert Rathke (ITT), and Umang Sharma (Desiccant Rotors/Bry-Air) for their assistance in providing pictures and diagrams.

Finally, I would like to thank my wife Kshani and sons Michael and Andrew for putting up with me over the last few years, when I spent most of my free time preparing the book.

DR. LAL JAYAMAHA

Abbreviations and Terms Used

AHU	Air handling unit
ASHRAE	American Society of Heating, Refrigerating and Air-conditioning Engineers
BAS	Building automation system
CAV	Constant air volume
CO	Carbon monoxide concentration
CO_2	Carbon dioxide concentration
ESM	Energy saving measure
FCU	Fan coil unit
kW	Kilowatt
kW/RT	Measure of chiller efficiency
kWh	Kilowatt hour
lm	Lumen
O_2	Oxygen concentration
OTTV	Overall thermal transfer value
Pa	Pascal
RH	Relative humidity
RT	Refrigeration tons
SC	Shading coefficient
SHGC	Solar heat gain coefficient
USgpm	US gallons per minute
VAV	Variable air volume
VSD	Variable speed drive
ΔP	Pressure differential
ΔT	Temperature differential

Energy-Efficient
Building Systems

Energy Management and Energy Auditing

1.1 Introduction

Due to economic and environmental reasons, organizations around the world are constantly under pressure to reduce energy consumption. As energy cost is one of the main cost drivers for businesses, reduction in energy consumption leads to reduction in operating costs, and thereby helps to improve the profitability of organizations.

One of the main environmental concerns relating to energy consumption is the emission of carbon dioxide (CO_2), which is a "greenhouse gas" that contributes to global warming. Due to the release of CO_2 during burning of fossil fuels, CO_2 emissions can be closely correlated to energy use.

Another environment-related concern is the ever-increasing demand for fossil fuels, such as oil and gas, to support economic development. Since they are nonrenewable energy sources that take millions of years to form, they draw on finite resources, which will eventually be depleted.

Reduction in energy consumption can be achieved through energy efficiency and energy conservation programs. Such programs involve the promotion of efficient or effective use of energy, which helps to save energy and results in reduced environmental pollution and operating costs.

The first chapter of the book provides an overview of how such energy saving programs can be formulated through energy management and energy auditing. Thereafter, the subsequent chapters of the book cover, in detail, energy saving opportunities associated with different building systems and how to identify them, and the implementation of suitable solutions.

1.2 Energy Management

Energy management is a procedure for containing and reducing the overall energy consumption and energy costs of an organization. Some typical objectives of energy management, which depend on the needs of each individual organization, include; lowering operating cost, increasing profitability, reducing environmental pollution and improving working conditions.

For an energy management program to be successful, it needs the commitment and support of the organization's management and should be in synergy with the organization's objectives.

Energy management requires a systematic approach— from the formation of a suitable team, to achieving and maintaining energy savings. A typical process is outlined in Fig. 1.1.

The first step is to select a team with the right skills to lead and execute the energy management program. In many cases, the organization may not have people with the necessary technical expertise to implement

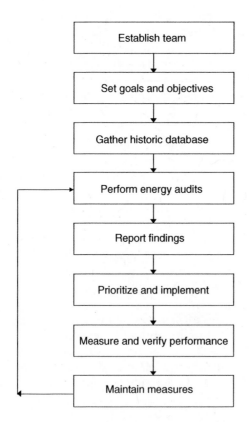

Figure 1.1 Typical energy management program.

all the aspects of an energy management program. In such situations, a professional energy management company could be engaged to carry out some of the tasks, but the organization should appoint a suitable person or a team who will have the overall responsibility for the program.

The energy management team then needs to set the objectives and priorities for the energy management program. Some aspects that should be covered are targeted savings, available budgets, and time frame for the project.

It is also recommended to create a historic database so that energy savings achieved in the future can be evaluated by comparing with this database. The historic database, depending on the type of facility, should include data such as utility consumption, occupancy rates, operating hours, occupied building floor areas, and production volumes.

Next, specific energy saving measures (ESMs) need to be identified. This is the most important part of the energy management program, and is achieved through *energy audits*. Energy audits can involve different levels of detail, depending on the objectives of the study. As described later (section 1.3), the American Society of Heating, Refrigerating and Air-conditioning Engineers (ASHRAE) has categorized audits into three levels, from 1 to 3, depending on the depth of the audit. For example, if the energy management team is considering which buildings, out of a group of buildings, have the best potential for savings, then a Level I preliminary walk through audit may be sufficient. On the other hand, if the energy management team requires an estimate of potential savings and cost for a particular facility for financial forecasting, a Level II energy survey and analysis may be necessary. Similarly, if the objective is to identify and implement specific savings measures, a Level III audit may be required.

Once the audit is completed, the reported findings should be used to prioritize the implementation of the ESMs. Thereafter, the ESMs should be implemented and the postinstallation performance data monitored to ensure that projected savings are achieved. In the event that savings fall short of the targeted values, remedial action needs to be taken.

Finally, once all the measures are implemented, they need to be maintained to ensure that the achieved savings can be sustained over the expected life of the system. Where ever necessary, follow-up energy audits should be performed periodically.

As can be seen, the most important part of an energy management program is an energy audit to identify potential energy savings measures. The next section of this chapter covers the different aspects of energy audits.

1.3 Energy Audits

Energy audits are carried out to understand the energy performance of buildings and facilities so that areas with potential for energy savings can be identified. An energy audit consists of a study of how a building or facility uses energy, how much it pays for the energy, and the identification and recommendation of improvement measures to reduce energy consumption.

The scope of work undertaken in an energy audit depends on the objectives of the study and resources available. As per the ASHRAE application handbook, energy audits can be classified into three main categories, based on the scope of work covered in the study. These three categories are described next.

1.3.1 Level I—Walk-through assessment

This involves the assessment of a building's energy cost and efficiency through the analysis of energy bills and a brief survey of the building. A Level I energy survey helps to identify and provide savings and cost analysis of low-cost or no-cost measures. It also provides a listing of potential capital improvements that merit further consideration, along with an initial judgment of potential costs and savings. The level of detail depends on the experience of the person performing the audit and on the specifications of the client paying for the audit.

Walk through studies provide an initial assessment of savings potential for buildings and therefore help to optimize available resources by being able to identify buildings with the best potential for savings and where further effort and study should be devoted.

1.3.2 Level II—Energy survey and analysis

Level II audits include a more detailed building survey and energy analysis. A Level II energy audit identifies and provides the savings and cost analysis of all practical measures. It also provides a listing of potential capital intensive improvements that require more thorough data collection and analysis, along with an initial judgment of potential costs and savings.

Level II audits normally do not include data logging, but may involve some "spot measurement" of parameters such as motor power, space temperature and relative humidity, and airflow rates, where necessary.

Although Level II studies require more resources than Level I studies, they are able to identify and short list not only buildings for further study (as in Level I surveys) but also areas or measures within a building that have a good potential for savings and where further study should be carried out. Therefore, Level II studies are useful exercises to be carried out before a detailed study so that the resources available for the detailed study can be better utilized.

Following are some of the main tasks carried out during a Level II audit:

Collection of information on facility operations. Since the auditor should have a good understanding of the facility, a site plan and a set of building plans, showing the layouts of the different areas of the facility, are reviewed. Other information, such as building floor areas, operating hours for the facility, and details of equipment used, with technical specifications, also need to be collected.

Utility bill analysis. Past utility consumption data for at least a one year period is collected and plotted (Fig. 1.2). The data should ideally be for a calendar year so that seasonal fluctuations can be identified.

The data can then be used to estimate monthly and annual indices for kWh/ft^2 and \$/ft^2 which can be used to compare the energy efficiency of the building with other buildings having the same characteristics.

Two commonly used indices are annual energy cost per square foot and annual kWh use per square foot. Some typical data plots are shown in Figs. 1.3 and 1.4.

Another electricity consumption related index is the *load factor*.

$$\text{Load factor} = \frac{\text{total monthly kWh consumption}}{\text{maximum kW demand} \times 24 \text{ hours} \times \text{no. of days in month}}$$

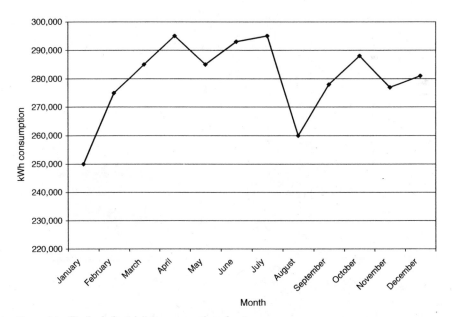

Figure 1.2 Typical electricity consumption chart.

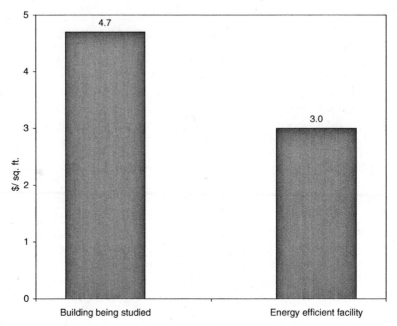

Figure 1.3 Annual energy cost per square foot.

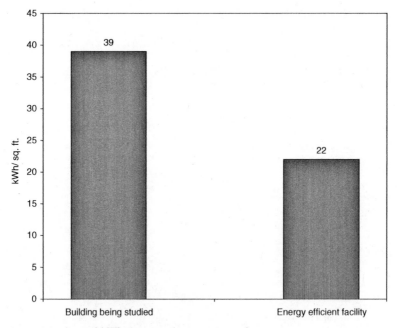

Figure 1.4 Annual kWh consumption per square foot.

Load factor is the relationship of the facility's kWh consumption to its monthly peak demand. Profiling these data by month can indicate how and when a facility consumes electrical energy, if equipment is being shut down at night, and how much of the building's demand profile is weather sensitive.

End user profile. Understanding where and when energy is being consumed is an important first step to understanding where it can be saved. The end use profile, which breaks down the total consumption into the different end users, helps to define where money is being spent and to focus efforts where the highest net financial gain can be obtained.

The actual end user breakdown will depend on the facility and the different systems it contains. For a typical commercial building, the consumption can be grouped into the following:

- Chillers
- Boilers
- Pumps
- Cooling towers
- Air handling units (AHUs)
- Fan coil units (FCUs)
- Ventilation fans
- Lighting
- Miscellaneous

The consumption for each system can be estimated using the individual kW consumption multiplied by the annual operating hours, as illustrated in Example 1.1.

Example 1.1 Consider a chiller of capacity 500RT, which operates 55 hours a week at an average loading of 70 percent. If the efficiency of the chiller at this loading is 0.65 kW/RT, the energy consumption of the chiller can be estimated as follows:

$$\text{kWh consumption} = 500 \text{ RT} \times 0.65 \text{ kW/RT} \times 70\% \times 55 \text{ hours/week} \times 52 \text{ weeks/year} = 650,650 \text{ kWh/year}.$$

Similarly, consumption for other equipment and systems such as pumps, fans, and lighting can also be computed.

The estimated individual consumption values should add up to the total metered consumption from utility bills. If the total of the estimated values do not add up to the actual consumption figure, it indicates that all the users have not been accounted for or some of the users may not have been correctly estimated. In such a situation, further investigation

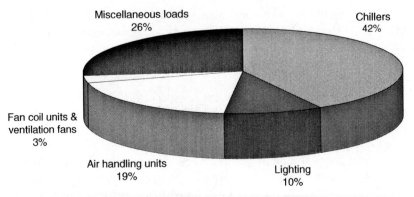

Figure 1.5 Typical end use consumption chart.

is necessary to ensure that all users are accounted for. The end use consumption can be plotted as a pie-chart, as shown in Fig. 1.5.

Comparison with benchmarks.   The computed values of kWh consumption, kW demand, and cooling load can be used to assess how efficient the facility is by comparing with benchmarks from other buildings.

Some benchmarking parameters and what they indicate are given in Table 1.1.

For industrial plants, the specific energy consumption, which relates the energy consumed in kWh of electricity or liters of fuel to the product output, can be computed. These specific energy consumption values can then be compared with other plants manufacturing identical or similar products.

Identify and shortlist areas.   The above benchmarking exercise assists in short-listing areas that have the potential for energy saving and, therefore, helps to optimize the use of resources.

TABLE 1.1 Some Typical Benchmarking Parameters for Buildings

Parameter	Description	What it means
W/ft^2 (lighting)	Lighting wattage/floor area served	Efficiency of lighting systems
RT/ft^2	Cooling load/air-conditioned floor area	Effectiveness of the building envelope
kW/RT (chiller/s)	Chiller kW/cooling load	Efficiency of chiller/s
kW/RT (chiller system)	(Total kW for chillers, pumps & cooling towers)/ cooling load	Efficiency of entire chiller system

For example, if the lighting efficiency for a building is found to be 2 W/ft^2 as compared to the benchmark of 1 W/ft^2 for buildings of similar nature, it indicates that the possibility of achieving significant energy savings from lighting for that building is good.

An estimate of potential savings can then be made using the current and proposed benchmark values (Example 1.2).

> **Example 1.2** Consider a building with a cooling load of 1000 RT. The chiller system operates, on an average, 55 hours a week and the efficiency of the chiller system is 0.9 kW/RT. If based on benchmarking data, the system efficiency can be improved to 0.7 kW/RT,
>
> $$\text{Estimated savings} = 1000 \text{ RT} \times (0.9 - 0.7) \text{ kW/RT} \times 55 \text{ hours/week}$$
> $$\times 52 \text{ weeks/year} = 572{,}000 \text{ kWh/year}$$

Similarly, savings can be computed for other areas of the facility. These estimated savings values and associated costs can then be used to identify whether a particular building, or which system within a building, has the potential for savings before proceeding to a Level III audit.

1.3.3   Level III—detailed energy audits

Level III audits focus on potential optimization and capital intensive projects identified or short-listed during Level II audits and involve more detailed field data gathering and engineering analysis. They also provide detailed project cost and savings information with a high level of confidence, sufficient for major capital investment decisions. Therefore, Level III studies are also sometimes called *investment grade audits* (IGA).

Following are some of the main tasks carried out during a Level III audit:

Introductory meeting. The first step in initiating a detailed audit is the kick off meeting with the facility management team. The meeting normally involves key members of the facility management team, including the facility manager and maintenance supervisors. Some of the aspects usually covered at the kick off meeting are:

- Introduction of audit team to the facility management team
- Procedures for reporting and transmitting all correspondence and documentation
- Preliminary work schedule
- Special work considerations (if any)
- Safety issues

- Customer review and approval process
- Schedule for future progress meetings
- Access to information, drawings, utility data, metering data
- Site access

Audit interviews.   The next step is to interview or meet the relevant people to gather accurate information that are required for the audit.

The facility manager, or any other relevant official, needs to be interviewed to collect information such as equipment specifications, drawings, operational data, and utility data.

The maintenance supervisor also needs to be consulted to gather information on equipment operation strategies, equipment performance, process problems, and other details such as types of lamps and ballast used.

Data collection and logging.   Data collection and logging is the most important part of the detailed study, where data on equipment and operations is collected so that ESMs can be identified.

Depending on the facility, data collection can cover areas such as, air-conditioning systems, lighting, ventilation systems, motors, water heaters, boilers, compressed air systems, and other specialized equipment.

Commonly used instruments

Portable power meters help measure power consumption, current drawn, and power factor. The meters should have a clamp-on feature to measure current and probes to gauge voltage so that measurements can be recorded without any disruption to normal operations.

Portable light meters measure lighting levels. Normally, light meters measure in footcandles or lux.

Tachometers are useful for measuring the speed of pumps and fans. Optical type tachometers are preferable due to the ease of measurement afforded by them.

Airflow measurements can be made using anemometers, which can measure the velocity of air flowing in ducts and at openings. The two types of anemometers available are "vane" and "hot-wire." The measured air velocity and flow area can be used to compute the volume flow rate of air, usually in cubic feet per minute (cfm) or cubic metres per hour (cmh).

Airflow hoods are used to measure the volume of airflow through supply and return diffusers. Airflow hoods contain an air velocity integration manifold that averages the air velocity across a fixed flow area and gives the airflow volume.

Flow meters are used for measuring the flow rate of chilled water, hot water, or condenser water in pipes. Ideally, flow meters should be the portable clamp-on type, which can be installed without disrupting the flow in pipes. Portable flow meters are of the ultrasonic type, which use ultrasonic transmitter and receiver probes clamped onto the bare pipe surface (insulation has to be removed for chilled water and hot water pipes).

Temperature sensors are required to measure chilled water, hot water, condenser water, and air temperatures. The most commonly used sensors are RTDs and thermistors. The accuracy of these sensors is important for applications such as measuring the temperature difference between chilled water return and supply, to compute the cooling load. For such applications, specially calibrated sensors should be used. Such temperature sensors need to be connected to a data logger so that temperature data can be monitored and logged.

kW sensors, which have current transformers, potential transformers, and kW transducers, can be used for measuring the power consumption of major equipment like chillers. They normally come with clamp-on or split core current transformers that can be fitted onto the current-carrying conductors inside the electrical panel without the necessitating switching off of the power supply. Such devices usually give a milliamp or voltage output which needs to be connected to a data logger.

Pressure gauges, manometers, and *pressure transducers* are used for measuring the pressure in pipes and ducts. While pressure gauges are used for taking instantaneous pressure readings, pressure transducers are used where the profile of the pressure is required. Pressure transducers too need to be connected to a data logger.

Data loggers are used to monitor and log data such as temperatures, flows, motor current and power, and pressures. Data loggers are normally portable and can accept different inputs from sensors.

Combustion analyzers are portable devices that are used to estimate the combustion efficiency of boilers and other fossil fuel burning equipment by sampling the exhaust gases.

Other standalone portable loggers such as *amp loggers, run time loggers, temperature and* relative humidity *(RH) loggers,* and *lighting loggers* are also used depending on the need.

Data collection.   The parameters to be monitored vary from one facility to another due to differences in designs and operations. Following are some of the typical parameters that are measured in different systems:

- *Chillers*
  - Chiller motor power
  - Chilled water supply and return temperatures
  - Condenser water supply and return temperatures

- Flow rates
- Cooling load profile
- Chiller efficiency (kW/RT)

- *Cooling towers*
  - Fan power
  - Fan speed
  - Condenser water supply and return temperatures
  - Outside-air dry bulb/wet bulb temperatures

- *Pumps*
  - Pump motor power
  - Operating hours
  - Pressure across pumps
  - Flow rate
  - Distribution system pressures

- *Air handling units and air distribution systems*
  - Fan motor power
  - Fan speed
  - Operating hours
  - Static pressure across fan
  - Supply and return airflows
  - Off-coil temperature
  - Return air temperature and RH
  - Outside air temperature and RH
  - Chilled water supply and return temperatures

- *Lighting*
  - Lighting operating hours
  - Power consumed by lighting circuits
  - Lighting levels

- *Steam and hot water systems*
  - Makeup water flow
  - Makeup water temperature
  - Condensate return temperature
  - Feedwater temperature
  - Feedwater/steam flow
  - Flue gas composition and temperature
  - Steam pressure
  - Fuel usage
  - Boiler efficiency
  - Amount of condensate recovered

Data analysis.   Data collected are usually in raw form and need to be refined to enable analysis. Some of the steps involved in data analysis are listed here.

Computation of useful parameters.   Various parameters that are not directly measured need to be computed using the measured data. Some examples include cooling load, which is computed using chilled water flow and chilled water temperatures; chiller efficiency, which is computed using the actual cooling load and compressor motor power.

Tabulation of data.   Some data needs to be tabulated so that meaningful calculations and analysis can be performed (see Table 1.2).

Plotting of data.   As part of the data analysis, the measured data needs to be plotted so that various operating trends and characteristics can be identified. Some parameters are plotted against time to see how they vary. For example, Fig. 1.6 (case study described later), shows how a building's cooling load demand varies with time. Similarly, some parameters need to be plotted against other parameters to see how they interrelate, as in Fig.1.7, which shows how the efficiency of a chiller varies with loading.

Comparing performance with specifications.   The performance of equipment or systems also needs to be compared with design data so that shortcomings or possible improvements can be identified. Table 1.2 shows a typical comparison of measured data with design data for a group of chillers.

TABLE 1.2   Typical Comparison of Measured Data with Design Data for a Group of Chillers

Parameter	Design	Chiller 1	Chiller 2	Chiller 3
Chilled water supply temp (°C)	5.6	5.3–5.6	5.5–6	5.4–5.7
Chilled water return temp (°C)	12.2	9–10	9–10	9–10
Chilled water flow rate (GPM)	1500	1828–1875	1600–1,700	1550–1680
Condenser water supply temp (°C)	32.2	24–26	24.6–26	24.7–26.5
Condenser water return temp (°C)	37.5	28–31	29–31.5	28–30.5
Condenser water flow rate (GPM)	2250	2000	1550–1600	2070–2126
Capacity (RT)	750	490–610	400–550	450–575
Power (kW)	504	330–400	300–400	300–375
Efficiency (kW/RT)	0.672	0.65–0.70	0.70–0.75	0.65–0.75
Evaporator ΔP (bar)	0.41	0.63	0.54	0.57
Condenser ΔP (bar)	0.55	0.57	0.34	0.61
CW approach Temp(°C)	—	2.9	0.8	2.5

Following are some examples where the actual performance needs to be compared with design values:

- Chiller capacity and efficiency
- Boiler capacity and efficiency
- Chilled water flow
- Condenser water flow
- Chilled water supply and return temperatures
- Condenser water supply and return temperatures
- Cooling tower capacity and performance
- Pump flow, head and motor power
- AHU cooling capacity, supply and fresh airflow
- Motor power
- Ventilation rates

Identification of ESMs and concept design.   Once the data is analyzed, the ESMs can be identified. The ESMs identified vary from one facility to another due to differences in equipment, system design, and operations. Some of the common ESMs are listed below and explained in detail in the subsequent chapters of the book.

- Chillers
  - Providing design flow rates and temperatures
  - Reducing operating hours
  - Sequencing chillers
  - Resetting chilled water and condenser water temperatures
  - Consolidating the chiller plant
  - Using small chillers for night operations
  - Replacing inefficient chillers
  - Balancing water flow rates
  - Installing automatic condenser tube cleaning systems

- Cooling towers
  - Installing variable speed controls for cooling tower fans
  - Replacing under performing cooling towers
  - Reducing operating hours
  - Adding capacity if existing towers are undersized

- Pumps
  - Reducing capacity of pumps (trimming impeller, reducing speed, or replacing pumps)

- Converting constant speed pumps to variable speed pumps
- Optimizing pump operating strategies
- Removing unnecessary restrictions in piping systems
- Water flow balancing

- Air handling units and air distribution systems
  - Reducing fan speed/use of variable speed drives (VSDs)
  - Converting constant air volume (CAV) systems to variable air volume (VAV) systems
  - Improving controls and control strategies
  - Reducing operating hours
  - Ensuring ventilation control based on occupancy
  - Cleaning/replacement of coils and filters
  - Air balancing for distribution systems

- Lighting
  - Replacing inefficient lamps with high-efficiency lamps
  - Using high-efficiency ballasts
  - Reducing lighting levels
  - Installing timer controls
  - Installing occupancy sensors
  - Using day lighting

- Steam and hot water systems
  - Improving boiler operating efficiency
  - Heat recovery from flue gas
  - Heat recovery from blow down
  - Condensate recovery
  - Reduction of boiler pressure
  - Reduction of boiler operating hours

- Compressed air systems (not covered in this book)
  - Eliminating leaks
  - Reducing operating pressure
  - Increasing downstream storage capacity
  - Reducing inlet air temperature
  - Recovering heat from compressors
  - Improving compressor controls
  - Installing variable speed compressors

Cost and savings analysis.   Once the ESMs are designed, the cost for implementing each measure and the achievable savings are estimated.

Cost estimation is usually done by listing the scope of work necessary for each ESM and using unit rates for the different tasks (if available), or by obtaining quotations from suitable contractors.

Once the cost and savings for each measure is computed, the financial viability for each can be considered based on agreed financial criteria. The most common financial criteria used for evaluating energy management projects are, simple payback period, life-cycle cost, and internal rate of return (IRR).

Baseline data for savings verification.   Usually, during the audit, a baseline for system performance is also established so that postinstallation system performance can be compared with these preinstallation data to estimate the achieved savings. Such data are very important if the project is undertaken as a "performance contract," since the Energy Services Company (ESCO) will be paid based on savings achieved. Even if it is not a performance-based project, it is still a good practice to have baseline data so that postinstallation savings can be quantified and compared with projected savings.

1.4   Case Study

The following case study is used to illustrate typical findings and recommendations for an energy audit covering the central chilled water system of a building. As such, it should be noted that the audit covers only one system within the facility and that the identified energy saving measures may not be applicable to other buildings (typical energy saving measures applicable to different systems within buildings are explained in detail in the subsequent chapters of the book).

1.4.1   Description of the system

The central chilled water plant consists of six sets of chillers of 450 RT capacity, each with associated cooling towers, condenser water pumps, and primary and secondary chilled water pumps.

The central chilled water plant operates 24 hours a day and provides chilled water to an office tower, a residential apartment building, and a retail podium block. Chilled water is supplied to the office tower building and residential apartments 24 hours a day, while the retail areas are supplied daily with chilled water only from 9 a.m. to 10 p.m.

The number of chillers in operation is controlled based on the building's cooling load, and the building automation system (BAS) automatically cuts in and cuts out chillers, with their associated pumps and cooling towers, depending on the load.

1.4.2   Main findings and recommendations

Provide additional small chiller.   As can be seen from Fig. 1.6, the actual daytime cooling load (between 9 a.m. and 6 p.m.) varies from 1300 to

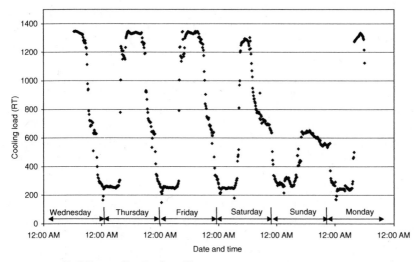

Figure 1.6 Building cooling load profile.

1350 RT on weekdays and on Saturday mornings. The cooling load on Saturdays, after about 1 p.m.) and on Sundays varies between 600 and 700 RT. The daily night load is between 200 and 250 RT.

The efficiency of chillers depends on how much they are loaded and Fig. 1.7 shows the loading versus efficiency for one of the chillers. Based

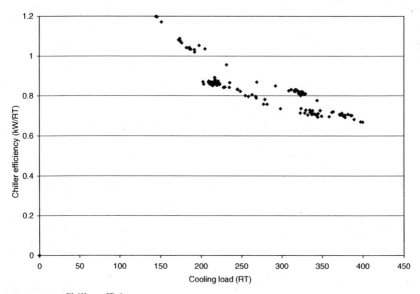

Figure 1.7 Chiller efficiency.

on the measured data, the actual operating efficiency of the chillers is between 0.62 and 1.2 kW/RT. The system efficiency (Fig. 1.8) for the chiller system, which includes the efficiency of the chillers as well as that of the chilled water pumps, condenser water pumps, and cooling towers, is 1.1 kW/RT during daytime. The system efficiency reduces to between 1.4 and 2.4 kW/RT during nighttime operation.

The prime reason for the drop in system efficiency at night is due to the need for running a 450-RT capacity chiller to meet the night load of 200 to 250 RT. During this time, the operating chiller is loaded only to 44 to 55 percent of its capacity. This results in a drop in chiller efficiency from 0.65 kW/RT to about 0.9 kW/RT.

Therefore, it was proposed to install an additional small chiller to act as the "night chiller." The proposed chiller capacity was 250 RT to meet the night load of 200 to 250 RT. This would help improve chiller efficiency in the night from 0.9 kW/RT to about 0.6 kW/RT.

Reduce capacity of primary chilled water pumps and condenser water pumps. The primary chilled water pumps and condenser water pumps are fixed speed pumps and one set each is operated with one chiller.

The pumps have constant flow valves installed to control the flow of water to each chiller. As the capacity of the pumps is much higher than

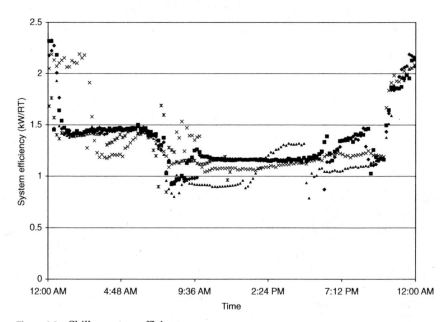

Figure 1.8 Chiller system efficiency.

required for the application, these constant flow valves reduce the water flow by inducing extra pressure drop in the system. Further, the condenser water flow through some chillers was about 20 percent more than required.

Therefore, it was recommended to remove the constant flow valves that induce high-pressure losses in the system, and to reduce the capacity of the pumps by trimming the impellers to suit the actual system requirements.

Secondary chilled water pumps.   The secondary chilled water pumps (constant speed pumps) are arranged in three groups—the office tower, residential apartments, and the retail area. Each group of pumps has three pumps installed in parallel and a maximum of two pumps are operated to meet the peak load, while the remaining pump acts as a standby. There are also throttling valves installed at the discharge of each pump to reduce the water flow by inducing extra pressure drop in the system.

The chilled water system is designed to operate at a temperature difference (ΔT) of 5.6°C between chilled water return and supply. In Fig. 1.9, the actual chilled water ΔT during daytime for the office tower is 5°C to 6°C but drops to about 2°C to 3°C during nighttime and during

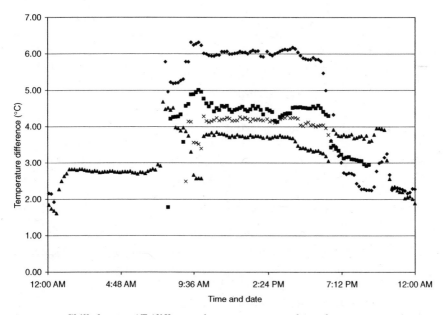

Figure 1.9   Chilled water ΔT (difference between return and supply temperatures).

daytime on Sundays. The chilled water ΔT for the retail area varies from 4°C to 5°C, while that for the residential apartments varies from 3°C to 4°C during daytime and 2°C to 3°C at night. This indicates that there is significant overpumping of chilled water to the residential apartments at all times and to the office tower at night.

Therefore, it was proposed to convert the existing constant speed secondary pumping systems to variable speed pumping systems. Each group of three pumps serving a particular area of the facility would have a differential pressure sensor, which can be used to maintain a set pressure in that chilled water system by varying the speed of the pumps.

Operation of chillers.   The temperature of chilled water returning from the residential apartments is lower than that returning from the office tower and retail areas. The chilled water return pipe from the residential apartments is connected to the main chilled water return header in the chiller plant room, which is closer to chillers 1 and 2, while the return pipe from the office tower is connected near chillers 5 and 6. Therefore, during normal operation, the temperature of return chilled water to chillers 5 and 6 is higher than that to the other chillers. Since the chilled water flow rate and the chilled water supply temperature are fixed for each chiller and the loading of each chiller depends on the return chilled water temperature (higher return temperature leads to higher loading of chillers), chillers 5 and 6 are loaded more than the other chillers. Therefore, during daytime, when four chillers are in operation, chillers 1 to 4 cannot be fully loaded. This leads to situations where four chillers (450 RT $\times$ 4 = 1800 RT) need to be operated to meet a cooling load of less than 1350 RT, when only three chillers (450 RT $\times$ 3 = 1350 RT) are necessary (Fig.1.10).

Therefore, when the operations of the secondary chilled water pumps are optimized by converting from constant flow to variable flow systems, it will not only result in pumping energy savings, as described earlier, but will also lead to better chiller loading (higher chilled water return temperature) and further energy savings from improved chiller efficiency. In addition, the small chiller proposed for nighttime operation can also be used during the daytime to act as a swing chiller to supplement the big chillers and better match operating chiller capacity to the cooling load.

As illustrated in this case study, the data collection and analysis carried out as part of an energy audit can help identify shortcomings in various energy consuming systems so that suitable solutions can be recommended to overcome them.

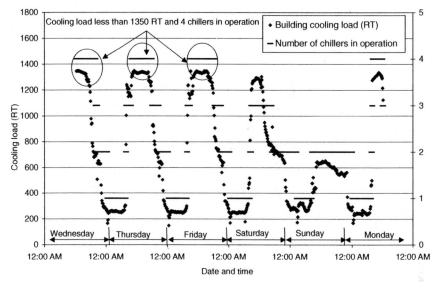

Figure 1.10 Cooling load and number of chillers in operation.

1.5 Summary

The chapter provided an overview of energy management and energy audit programs and highlighted how they are useful for any building, organization, or industry to reduce energy consumption. The key tasks of energy audits, which provide a systematic approach for identifying energy saving opportunities, were described in detail. Typical findings and recommendations of energy audits were also illustrated using a simple case study.

The subsequent chapters of this book describe, in detail, the different energy consuming systems in buildings and how suitable energy saving measures or solutions can be applied to reduce the overall energy consumption of buildings.

Air-Conditioning and Central Chiller Systems

2.1 Introduction

Air-conditioning systems used in commercial and institutional buildings can account for more than 50 percent of the total electricity consumed. However, they are essential for buildings in hot and humid climates, where air-conditioning is used to provide a comfortable internal environment for the occupants so that they can work and perform productively. Therefore, it is important to understand how air-conditioning systems work so that they can be made to operate as efficiently as possible.

2.1.1 Refrigeration

Air-conditioning is normally used to remove heat from an occupied space and maintain it at a temperature lower than the outdoor temperature. This requires the use of a refrigeration system that can help maintain a body at a comfortable temperature.

As shown in Fig. 2.1, a refrigeration machine "R" absorbs heat from the cold body at temperature T_o and releases the heat Q_s into the surroundings at temperature T_s. This process requires work "W" to be done on the system. The heat released into the surroundings equals the heat absorbed from the cold body plus the work done or mechanical energy consumed.

Refrigerants are used in refrigeration machines to perform this function. Refrigerants evaporate at low temperatures and condense at higher temperatures. Therefore, refrigerants can be evaporated at low temperatures, during which heat is absorbed, and then condensed at a higher temperature and pressure, when the absorbed heat (plus that due to the work done) is released into the atmosphere.

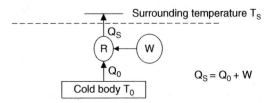

Figure 2.1 Operating principle of a refrigeration machine.

2.1.2 Vapor compression cycle

The vapor compression cycle is the most commonly used refrigeration cycle. It has four mechanical components through which the refrigerant is circulated in a closed loop, as shown in Fig. 2.2.

The *refrigerant* enters the compressor as low-pressure vapor and is compressed to high-pressure vapor. The high-pressure vapor then flows to the condenser, which is a heat-exchanger, where heat is released from the refrigerant and the refrigerant condenses from high-pressure vapor to high-pressure liquid. Next, the high-pressure liquid refrigerant flows through the expansion valve to the evaporator and becomes low-pressure liquid refrigerant. In the evaporator, the liquid refrigerant evaporates at low temperature by absorbing heat from the surroundings. This is the cooling effect of the refrigeration system. The refrigerant then becomes low-pressure vapor and enters the compressor, and the cycle is repeated.

The vapor compression cycle can be represented on a pressure–enthalpy (p-h) diagram, as shown in Fig. 2.3, for an ideal case without losses.

2.1.3 Absorption cycle

Another refrigeration cycle is the absorption cycle. This cycle is heat driven unlike the vapor compression cycle, which is work driven. The

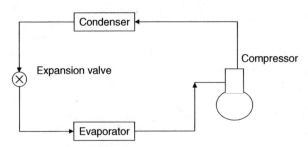

Figure 2.2 Vapor compression cycle.

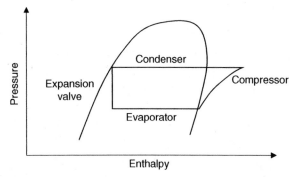

Figure 2.3 Pressure–enthalpy diagram for an ideal vapor compression cycle.

cycle involves a mixture of two substances, such as ammonia and water or lithium bromide and water, that attract each other.

Figure 2.4 shows the components of an absorption cycle. In this cycle, the solution of a salt, such as lithium bromide and water, enters the generator where heat is applied. The heat breaks the bonds of the mixture and evaporates some of the water. The water vapor goes to the condenser, where it is condensed from vapor to high-pressure liquid. This water, which acts as the refrigerant, passes through a metering valve and enters the evaporator as low-pressure liquid. The water then evaporates in the evaporator by absorbing heat, providing the cooling effect of the cycle. The low-pressure vapor then enters the absorber where the strong salt solution returning from the generator absorbs it and it is then pumped back as a weak solution into the generator to complete the cycle.

The main difference between the absorption cycle and the vapor compression cycle is that the former uses a heat source to power the generator as compared to doing "work" in the compressor of the vapor compression cycle. The absorption cycle is also less efficient than the vapor compression cycle. However, it is useful for applications where waste heat or cheap sources of heat are available.

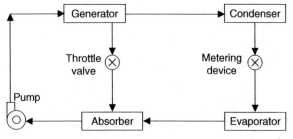

Figure 2.4 Components of the absorption cycle.

2.1.4 Refrigeration capacity

The capacity of air-conditioning machines is usually classified by how many "tons" of refrigeration (RT) effect they can produce.

1 RT of refrigeration is based on the amount of cooling 1 ton (2000 lb) of ice can produce by melting over a 24-hour period. 1 RT of refrigeration has the ability to remove 288,000 Btu in a 24-hour period.

$$1 \text{ RT} = 2000 \text{ lb} \times 144 \text{ Btu/lb} = 288,000 \text{ Btus}$$

Therefore, 1 RT = 12,000 Btu/h

Also, 1 RT = 3.517 kW.

2.1.5 Central air-conditioning systems

Buildings use different air-conditioning systems such as central systems, stand-alone package units, variable refrigerant volume (VRV) systems, and water-cooled package units. Out of these systems, the most commonly used system in large buildings is the central type. Therefore, this chapter will only cover central air-conditioning systems. In systems other than central ones, energy management strategies that can be adopted are usually specific to each installation and cannot be easily discussed in general.

In central air-conditioning systems, such as those found in commercial buildings, the evaporator of a chiller cools water. This chilled water is then pumped to air handling units (AHUs) and fan coil units (FCUs) located in different areas of the building. The AHUs and FCUs have fans that blow air through heat exchanger coils to transfer heat from the air to the chilled water, thereby cooling the air. The chilled water, after picking up heat from the coils, is then pumped back to the evaporator of the chiller to cool down and complete the cycle. Such systems are called chilled water systems. A typical arrangement of a central chilled water system is shown in Fig. 2.5.

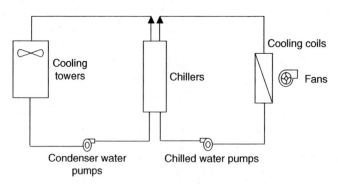

Figure 2.5 Typical arrangement of a central chilled water system.

Chillers used in central chilled water systems normally work on the vapor compression cycle or the absorption cycle. Since electric chillers working on the vapor compression cycle are the most common, and significant energy savings can be achieved from these machines through various energy management strategies, the rest of this chapter is mainly devoted to such chillers.

A chiller working on the vapor compression cycle contains four main components—the compressor, evaporator, condenser, and expansion device. The heat absorbed in the evaporator and the heat added during compression is rejected in the condenser. In air-cooled chillers, heat from the condenser is released directly into the ambient air, while in water-cooled systems, heat is released into the condenser water, which is then circulated to cooling towers where the heat is finally released.

2.2 Chillers

2.2.1 Introduction

Chillers are the biggest energy consumers in central air-conditioning systems. In commercial buildings where air-conditioning systems account for more than half the total electricity consumption, chillers end up being the single biggest consumers. Therefore, their efficiency has a significant effect on the overall energy performance of these buildings.

Chillers are classified according to the type of compressor used for compressing the refrigerant. The main types of compressors used are reciprocating, scroll, screw, and centrifugal.

Reciprocating compressors use pistons and connecting rods driven by a crankshaft, as shown in Fig. 2.6. The crankshaft is driven by a motor. In open type drives, the drive motor is mounted externally and the crankshaft extends through a seal out of the crankcase. In hermetic systems, the motor and the compressor are contained in the same housing and the crankshaft and motor are in direct contact with the refrigerant. Normally, reciprocating compressors have a number of cylinders in one unit. For high-capacity applications, multiple compressors are used.

Scroll compressors use two interfitting, spiral-shaped scroll members (Fig. 2.7). One scroll rotates while the other remains stationary. Due to the profile of the scrolls, refrigerant drawn in through the inlet port is compressed between the scrolls during rotation and then released at the discharge port. Scroll compressors are used for low-capacity applications of up to about 50 kW.

Screw compressors are used for medium-capacity applications up to about 1055 kW (300 RT). They consist of a set of male and female helically

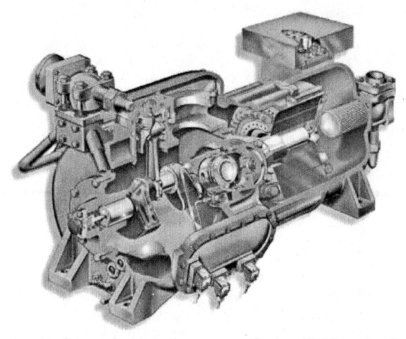

Figure 2.6   Cutaway of a reciprocating compressor. (*Courtesy of York International.*)

grooved rotors, as shown in Fig. 2.8. Compression is achieved by direct volume reduction due to rotation of the rotors. Refrigerant is taken in at the inlet port, compressed during rotation of the rotors, and finally released at the discharge port.

Centrifugal compressors consist of a single impeller (Fig. 2.9) or a number of impellers mounted on a shaft and rotating at high speed inside a housing. Refrigerant enters the impeller in the axial direction and is discharged radially at high velocity. The velocity pressure is then converted to static pressure in the diffuser. Centrifugal compressors are used for high-capacity applications, usually above 1055 kW (300 RT).

2.2.2   Chiller efficiency rating

Chiller efficiency is measured in terms of how many units of power is used to produce one unit of cooling.

$$\text{Chiller efficiency} = \frac{\text{Power input}}{\text{Cooling produced}}$$

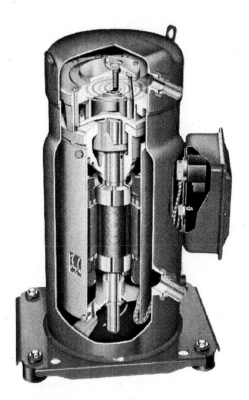

Figure 2.7 Cutaway of a scroll compressor. (*Courtesy of York International.*)

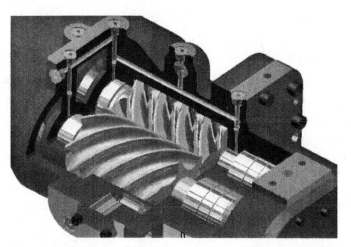

Figure 2.8 Cutaway of a screw compressor. (*Courtesy of York International.*)

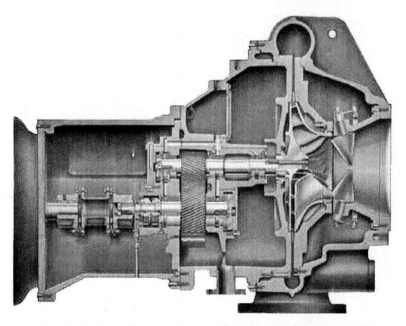

Figure 2.9 Typical centrifugal compressor. (*Courtesy of York International.*)

In the English IP system, chiller efficiency is measured in kW/RT

$$\text{where kW/RT} = \frac{\text{kW input}}{\text{Tons of refrigeration}}$$

In the SI metric system, chiller efficiency is measured in coefficient of performance (COP)

$$\text{where COP} = \frac{\text{kW refrigeration effect}}{\text{kW input}}$$

kW refrigeration effect = RT × 3.517

The energy efficiency ratio (EER), which is the ratio of the cooling capacity (Btu/h) to the electric power input (W), is sometimes used to rate reciprocating and scroll compressors in air-cooled chillers and direct-expansion refrigeration units.

Since chiller efficiency varies with loading, it is usually rated at full-load (100 percent of rated capacity) and part-load conditions (90, 80, 70 percent, and so on.). This information, normally provided by chiller manufacturers, is useful in selecting chillers for different applications.

Consider chiller A, which has an efficiency of 0.55 kW/RT at full load and 0.65 kW/RT at 70 percent load, and another chiller B, which has a full-load efficiency of 0.57 kW/RT and 0.61 kW/RT at 70 percent load. If based on the expected load profile of the building, the chiller is going to operate only for a short period of time at 100 percent load and most of the time at 70 percent load, then chiller B would be a better option due to its superior performance at the operating point that will be most frequently encountered. As will be explained later, to ensure optimum efficiency, the cooling load profile should be used to select the best chiller for the job based on chiller part-load data.

If the actual cooling load profile is not known, and the chiller is expected to operate under normally encountered conditions, the IPLV rating can be used to evaluate the chiller's performance. The IPLV rating calculates the chiller efficiency based on weightage given to efficiency at different points, which, in turn, is based on commonly encountered operating conditions.

$$\text{IPLV} = \frac{1}{\left[\left(\dfrac{0.17}{A}\right) + \left(\dfrac{0.39}{B}\right) + \left(\dfrac{0.33}{C}\right) + \left(\dfrac{0.11}{D}\right)\right]}$$

where A = kW/RT at 100 percent load
B = kW/RT at 75 percent load
C = kW/RT at 50 percent load
D = kW/RT at 25 percent load

2.2.3 Economizers

Some chillers use economizers that enable flash refrigerant gas to be introduced at intermediate pressure, between the evaporator and condenser pressures, into multistage compressors to improve efficiency.

Figures 2.10 and 2.11 show the arrangement of a two-stage economizer and the related pressure–enthalpy (p-h) diagram, respectively.

In this cycle, when refrigerant leaving the condenser passes through an orifice, some refrigerant is preflashed. The preflashing of some liquid refrigerant cools the remaining liquid. The refrigerant vapor is removed to ensure that the refrigerant at the economizer is in the liquid state (not saturated liquid and gas). As a result, the condition of refrigerant entering the evaporator has a lower enthalpy (moves further left on the p-h diagram) and increases the evaporator's refrigeration effect.

In chillers using 3-stage compressors, the flashing of refrigerant can be done at two different stages, and introduced between the first and second stages and second and third stages to further improve the efficiency of the chiller.

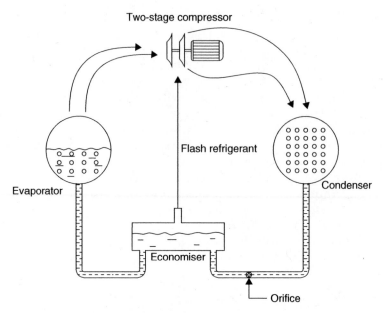

Figure 2.10   Arrangement of economizer.

2.2.4   Condenser subcooling

Chillers that do not have multistage compressors use liquid subcooling to improve the refrigerant's effect and, therefore, the chiller efficiency. In such a system, the refrigerant liquid leaving the condenser is further cooled (subcooled). This enables the refrigerant to enter the evaporator at a lower enthalpy, as shown in Fig. 2.12, to help increase the refrigerant's effect.

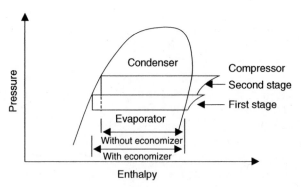

Figure 2.11   Pressure–enthalpy diagram for cycle with economizer.

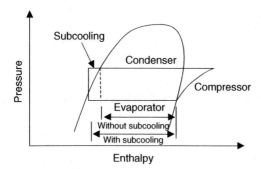

Figure 2.12 Pressure–enthalpy diagram for cycle with condenser sub cooling.

In the following sections of the chapter, various energy-saving measures for chiller systems are described.

2.3   Energy-Saving Measures for Chiller Systems

2.3.1   Water-cooled versus air-cooled chillers

The heat removed by the refrigeration cycle of an air-conditioning system is released in the condenser to either water or ambient air during condensing. Chillers are classified according to whether the condenser is water-cooled or air-cooled.

In air-cooled chillers, a single fan or a number of fans are used to blow ambient air through the condenser, which is normally a finned heat exchanger. Water-cooled chillers, on the other hand, have shell and tube heat exchangers, where the heat is released into the condenser water. This warm condenser water is then pumped to cooling towers where the heat is released into the environment. Heat transfer at the cooling towers takes place mainly by latent cooling, wherein some of the warm condenser water evaporates, absorbing the latent heat of evaporation from the condenser water, thereby cooling it (the operation of cooling towers is explained in Chapter 5). The condenser water, after releasing heat in the cooling towers, is pumped back through the condenser of the chiller to continue the heat removal cycle. However, since some of the water evaporates in the cooling towers during this cycle, "makeup water" needs to be constantly added to maintain the flow of condenser water through the system.

Water-cooled chillers operate at a lower condensing pressure than air-cooled chillers. The lower condensing pressure is due to releasing heat into the condenser water, which is first cooled to a temperature a few degrees above the wet-bulb temperature of the ambient air (by the cooling towers). In air-cooled chillers, heat is directly released to the ambient air and the heat transfer is dependent on the dry-bulb temperature of the air. Further, the heat transfer in the water-cooled shell and tube heat exchangers is better than in finned type air-cooled condensers.

The lower condensing pressure leads to a lower pressure differential between the evaporator and condenser, which results in lower power consumption by the compressor. Therefore, water-cooled chillers are far more efficient than air-cooled chillers.

Although air-cooled chillers are less efficient, they have many applications since they do not need a constant supply of makeup water (for condenser water). They also do not need cooling towers and can be mounted outdoors, such as on the roof of a building. Therefore, air-cooled chillers are used where a constant source of water is not available or where water is scarce. Air-cooled chillers are also used where space is limited, such as when building space for installing water-cooled chillers or outdoor space for installing cooling towers is unavailable.

The capital cost for water-cooled systems is normally higher than for air-cooled systems since the former needs an enclosed plant room, cooling towers, condenser water pumps, and extra piping for the condenser water system. The cost of piping for the condenser water system can be substantial in cases where the plant room is located in the basement and the cooling towers are located on the roof of the building.

Although the first cost of water-cooled systems is higher (than air-cooled systems), due to their superior energy efficiency, the extra initial cost can be paid back in just a few years. Even in situations where air-cooled chillers are currently in use, usually conversion to a water-cooled system, if practically possible, will yield a good return on investment.

However, when evaluating the benefits of water-cooled systems, the additional costs associated with water-cooled systems for makeup water and water treatment have to be considered as illustrated in Example 2.1.

Example 2.1 Consider a building that operates from 8 a.m. to 10 p.m. The current hourly kWh consumption can be computed, as shown in Table 2.1, by multiplying the hourly building cooling load by the present chiller efficiency (air-cooled). If these data are not available, a kWh consumption meter can be used to compute the electricity consumption of the chiller. Next, using manufacturers' efficiency data, the kWh consumption for the proposed water-cooled chiller can be estimated.

The difference between present (3915 kWh) and proposed (1803 kWh) total daily kWh consumption gives the daily savings (2112 kWh).

Additional savings will also result from reduction in peak power demand if the peaks for power usage and cooling load coincide.

Peak demand savings

     = Peak cooling load (RT)

        × difference in air-cooled and water-cooled chiller efficiencies (kW/RT)

     = 300 × (1.2 − 0.57)

     = 189 kW

Some of the above savings are offset by the extra electricity consumed by the condenser water pumps and cooling towers of the water-cooled system and can be computed based on their rated motor power for the application. Alternatively, the

TABLE 2.1 Estimation of Energy Savings for Conversion from an Air-Cooled Chiller to a Water-Cooled Chiller

Hours of operation	Number of hours A	Cooling load (RT) B	Present chiller efficiency (kW/RT) C	Present kWh consump-tion $A \times B \times C$	Proposed chiller efficiency (kW/RT) D	Proposed kWh consump-tion $A \times B \times D$
0800–0900	1	200	1.3	260	0.6	120
0900–1000	1	200	1.3	260	0.6	120
1000–1100	1	225	1.3	293	0.59	133
1100–1200	1	225	1.3	293	0.59	133
1200–1300	1	250	1.25	313	0.58	145
1300–1400	1	300	1.2	360	0.57	171
1400–1500	1	250	1.25	313	0.58	145
1500–1600	1	250	1.25	313	0.58	145
1600–1700	1	250	1.25	313	0.58	145
1700–1800	1	200	1.3	260	0.6	120
1800–1900	1	200	1.3	260	0.6	120
1900–2000	1	200	1.3	260	0.6	120
2000–2100	1	150	1.4	210	0.62	93
2100–2200	1	150	1.4	210	0.62	93
				3915		1803

value can be estimated by using a "rule of thumb" value of 0.15 kW/RT for the condenser water pumps and cooling towers.

Similarly, the makeup water consumption can be estimated assuming the combined losses due to evaporation, drift, and blow down to be 1.5 percent of the total condenser water flow. In typical systems, which are designed for a temperature difference of 5.6°C (10°F), between condenser water leaving and entering the chiller, the condenser water flow is 0.19 L/s per RT (3 USgpm per RT). In such a system, the makeup water consumption will be 2.8×10^{-3} L/s per RT (0.19 × 1.5%) or 1.02×10^{-2} m^3/hr per RT.

The extra cost for makeup water and electricity consumption for the condenser water pumps and cooling towers can be computed for the same building's cooling load profile as shown in Table 2.2.

Therefore, the net kWh savings = (2112 − 458) = 1654 kWh/day.

Makeup water consumption = 31.11 m^3/day.

If the utility tariffs are:

Electricity usage = $0.10/kWh

Peak demand = $10/kW per month

Water usage = $1/m^3

The net annual savings (based on operating 365 days a year)

= [(1654 kWh/day × electricity tariff) − (31.11 m^3/day

× water tariff)] × days/year + [189 kW

× monthly demand charges × 12 months/year]

= [(1,654 × 0.1) − (31.11 × 1)] × 365 + [189 × 10 × 12]

= $71,700

TABLE 2.2   **Extra Water Consumption for Water-Cooled Chillers**

Hours of operation	Number of hours A	Cooling load (RT) B	Consumption for pumps & cooling towers (kWh) A × B × 0.15 kW/RT	Makeup water consumption (m^3) B × (1.02 × 10^{-2})
0800–0900	1	200	30	2.04
0900–1000	1	200	30	2.04
1000–1100	1	225	34	2.30
1100–1200	1	225	34	2.30
1200–1300	1	250	38	2.55
1300–1400	1	300	45	3.06
1400–1500	1	250	38	2.55
1500–1600	1	250	38	2.55
1600–1700	1	250	38	2.55
1700–1800	1	200	30	2.04
1800–1900	1	200	30	2.04
1900–2000	1	200	30	2.04
2000–2100	1	150	23	1.53
2100–2200	1	150	23	1.53
			458	31.11

2.3.2   Chiller efficiency and life-cycle cost

Water-cooled chillers having efficiencies of 0.9 kW/RT were common in the 1970s. The efficiency of these chillers has improved over the last 30 years and the average chiller efficiency has now improved to about 0.55 kW/RT (efficiency of high-efficiency chillers can be better than 0.5 kW/RT).

Due to their design, high-efficiency chillers tend to be more expensive than chillers of average efficiency. Although high-efficiency chillers cost more, since they have a lower running cost (due to lower energy consumption), the extra initial capital cost is usually paid back in a few years.

The financial benefits of installing high-efficiency chillers instead of "lower" efficiency chillers can be confirmed through a life-cycle analysis that takes into account the initial capital cost of the chillers as well as the operating cost over their expected life, which is usually 10 to 15 years. The operating cost, which needs to be accounted for in a complete life-cycle costing, should include the main cost components such as the energy cost, peak power demand cost (if applicable), refrigerant cost, and maintenance cost. In addition, if the pressure drop across the evaporators and condensers of different chiller options considered are different, then the energy consumption of the chilled water and condenser water pumps to overcome the extra resistance also needs to be considered (discussed in Chapter 4).

Examples 2.2 and 2.3 illustrate life-cycle cost comparison for a 500-RT capacity chiller for three different efficiencies—0.5 kW/RT (high efficiency), 0.55 kW/RT (average efficiency) and 0.65 kW/RT (low efficiency). For simplicity, only the first cost and operating energy costs of the chiller are considered in the examples.

Example 2.2 (Chiller loading and efficiency are fixed)   In this example, the life-cycle cost for chillers with three different rated efficiencies of 0.5, 0.55 and 0.65 kW/RT are estimated assuming that the chillers operate 10 hours a day and 250 days a year.

The chiller loading (500 RT) and chiller efficiency are assumed to be fixed. If the electricity cost is $0.10/kWh and the electricity cost is expected to escalate by 2 percent a year, the annual energy cost for operating the chiller can be computed as follows:

Energy cost = Cooling load × hours of operation × efficiency (kW/RT)

$\qquad$ × electricity tariff.

For the 0.5 kW/RT chiller, the energy cost for Year 1

$\qquad$ = 500 RT × 10 h/day × 250 days/year × 0.5 kW/RT × $0.10/kWh

$\qquad$ = $62,500

The energy cost for Year 2 with 2 percent escalation of tariff will be $63,750 ($62,500 × 1.02).

Similarly, the annual energy cost for operating each of the three chillers can be computed.

The life-cycle cost computed for the three chiller combinations based on the following first cost is given in Table 2.3.

■ $300,000 for 0.5 kW/RT chiller

■ $275,000 for 0.55 kW/RT chiller

■ $250,000 for 0.65 kW/RT chiller

(All costs are annual energy costs except for Year 0, which includes the capital cost of the chillers.)

This simple life-cycle costing shows that the first option of using a 0.5 kW/RT high-efficiency chiller has the lowest total cost over a 10-year period, and is therefore the best option of the three.

Example 2.3 (Variable chiller loading)   Normally, the loading of chillers vary with time due to changes in the cooling load. Although, the energy cost for chillers can be computed using the same equation as in Example 2.2, since the chiller loading is not constant, the efficiency of the chillers cannot be taken as a fixed value. Therefore, the chiller energy consumption has to be computed at different loads and then summed up to give the total consumption. The easiest way to do this is by discretizing the cooling load into suitable intervals (1-hour intervals in this example) and computing the consumption for each interval, which can be added to give the total daily consumption.

TABLE 2.3   Life-Cycle Cost for the Different Chillers (Assuming Their Loading and Efficiency are Fixed)

	Annual cost of chillers		
	0.5 kW/RT	0.55 kW/RT	0.65 kW/RT
Year 0	$300,000	$275,000	$250,000
Year 1	$62,500	$68,750	$81,250
Year 2	$63,750	$70,125	$82,875
Year 3	$65,025	$71,528	$84,533
Year 4	$66,326	$72,958	$86,223
Year 5	$67,652	$74,417	$87,948
Year 6	$69,005	$75,906	$89,707
Year 7	$70,385	$77,424	$91,501
Year 8	$71,793	$78,972	$93,331
Year 9	$73,229	$80,552	$95,197
Year 10	$74,693	$82,163	$97,101
Total	$984,358	$1,027,793	$1,139,665

In this example, the cooling load is expected to vary, as shown in column A of Table 2.4. The energy consumption for the three different chillers can be estimated as shown in Tables 2.4 to 2.6.

(Efficiency values at different load conditions given in column B are assumed for this chiller and the other two chillers, but can normally be obtained from chiller manufacturers.)

If the building operates 250 days/year at the same load profile, the annual electricity cost for the 0.5 kW/RT efficiency chiller

= 2258.5 kWh/day × 250 days/year × $0.10/kWh

= $56,463

TABLE 2.4   Energy Savings Estimation for Case with Variable Cooling Load for 0.50 kW/RT Efficiency Chiller

Time	Cooling load (RT) A	Chiller loading	Chiller efficiency (kW/RT) B	kWh consumption C = A × B
0800–0900	350	70%	0.59	206.5
0900–1000	375	75%	0.58	217.5
1000–1100	400	80%	0.55	220
1100–1200	450	90%	0.52	234
1200–1300	500	100%	0.5	250
1300–1400	500	100%	0.5	250
1400–1500	450	90%	0.52	234
1500–1600	400	80%	0.55	220
1600–1700	400	80%	0.55	220
1700–1800	350	70%	0.59	206.5
			Total	2258.5 kWh/day

TABLE 2.5   Energy Savings Estimation for Case with Variable Cooling Load for 0.55 kW/RT Efficiency Chiller

Time	Cooling load (RT) A	Chiller loading	Chiller efficiency (kW/RT) B1	kWh consumption C1 = A × B1
0800–0900	350	70%	0.64	224
0900–1000	375	75%	0.63	236
1000–1100	400	80%	0.6	240
1100–1200	450	90%	0.57	257
1200–1300	500	100%	0.55	275
1300–1400	500	100%	0.55	275
1400–1500	450	90%	0.57	257
1500–1600	400	80%	0.6	240
1600–1700	400	80%	0.6	240
1700–1800	350	70%	0.64	224
			Total	2467 kWh/day

If the load profile of the building varies on some days (Saturdays and Sundays), the kWh consumption has to be computed separately for each typical pattern experienced and then multiplied by the number of days in a year that this pattern is expected to be experienced by the building.

The annual electricity cost for the 0.55 kW/RT efficiency chiller

= 2467 kWh/day × 250 days/year × $0.10/kWh

= $61,675

TABLE 2.6   Energy Savings Estimation for Case with Variable Cooling Load for 0.65 kW/RT Efficiency Chiller

Time	Cooling load (RT) A	Chiller loading	Chiller efficiency (kW/RT) B2	kWh consumption C2 = A × B2
0800–0900	350	70%	0.74	259
0900–1000	375	75%	0.73	274
1000–1100	400	80%	0.7	280
1100–1200	450	90%	0.67	302
1200–1300	500	100%	0.65	325
1300–1400	500	100%	0.65	325
1400–1500	450	90%	0.67	302
1500–1600	400	80%	0.7	280
1600–1700	400	80%	0.7	280
1700–1800	350	70%	0.74	259
			Total	2885 kWh/day

The annual electricity cost for the 0.65 kW/RT efficiency chiller

= 2885 kWh/day × 250 days/year × $0.10/kWh

= $72,125

TABLE 2.7   Life-Cycle Cost for Chillers Operating at Variable Load

	0.5 kW/RT	0.55 kW/RT	0.65 kW/RT	
Year 0	$300,000	$275,000	$250,000	(First cost)
Year 1	$56,463	$61,675	$72,125	(Electricity cost
Year 2	$57,592	$62,909	$73,568	increasing
Year 3	$58,744	$64,167	$75,039	at 2%
Year 4	$59,918	$65,450	$76,540	annually)
Year 5	$61,117	$66,759	$78,070	
Year 6	$62,339	$68,094	$79,632	
Year 7	$63,586	$69,456	$81,224	
Year 8	$64,858	$70,845	$82,849	
Year 9	$66,155	$72,262	$84,506	
Year 10	$67,478	$73,707	$86,196	
Total	$918,249	$950,324	$1,039,749	

The life-cycle cost comparison for the three different chillers is given in Table 2.7.

This example too shows that the 0.5 kW/RT chiller has a lower life-cycle cost. Ideally, the cost for the peak power demand needs to be added. This will be the power drawn by the chiller at the time when the building experiences peak power demand. For instance, if the building experiences peak power demand between 1200 and 1300 hours, the power drawn by the three chiller options will be 250 kW (500 RT × 0.5 kW/RT), 275 kW (500 RT × 0.55 kW/RT), and 325 kW (500 RT × 0.65 kW/RT), respectively.

If the monthly demand charges are $10/kW, the monthly demand cost for the three options will be $30,000 (250 kW × $10/kW per month × 12 months), $33,000, and $39,000, respectively. This additional cost then needs to be added to the projected annual energy cost for the different options.

If the maintenance costs for the different chillers being considered are appreciably different as can be the case if different brands of chillers are considered, the annual operating cost should be the sum of energy, power demand, and maintenance costs expected for each year of operation.

2.3.3   Sizing and configuration

As explained earlier, the operating efficiency of chillers depends on their loading. A typical chiller loading versus efficiency relationship is shown in Fig. 2.13. As can be seen from this figure, chiller efficiency is best when operating in the range 60 to 100 percent of its capacity, while optimum efficiency is obtained at 80 percent loading (some chillers operate best at 100 percent).

Chillers are usually oversized for new building installations due to the unavailability of accurate load estimation tools, leading to the use of high safety factors in the design. This over sizing repeats even when the systems are retrofitted since replacement of chillers is often done on a one-to-one basis. Also, chillers are normally sized to meet the peak load without much consideration of the load profile. Since the cooling load of

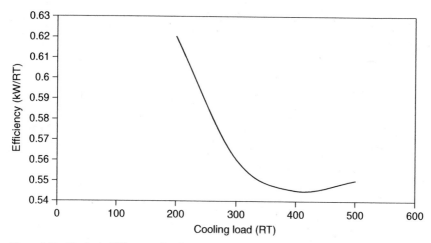

Figure 2.13 Typical chiller part-load curve.

a building varies with time, this can result in chillers operating at part load for long periods of the day, wasting much energy.

Therefore, sizing and configuration of chillers for new buildings should be done using accurate load estimation tools, while the measured cooling load profile should be used when retrofitting existing buildings. Chillers should be sized based on the peak cooling load and the cooling load profile of the building. If chillers of different capacities are available, depending on the load profile, various combinations of chillers can be operated during the day to match the building's cooling load profile. This would ensure that the chillers are able to operate within their best efficiency range at all times.

Examples 2.4 and 2.5 illustrate the importance of chiller sizing and configuration for a new chiller installation and a chiller plant retrofit, respectively.

Example 2.4 (new chiller installation) Consider the building cooling load profile shown in Fig. 2.14. The cooling load of the building varies from about 500 RT in the night to a peak of 1800 RT during the afternoon. This building cooling load profile can be satisfied by an infinite number of chiller combinations. If one chiller is used, an 1800-RT capacity chiller would be required. However, this would lead to very inefficient operation of the chiller during periods of low cooling load, (in the night when the load is about 500 RT). If two equally sized chillers of 900 RT each are used to meet the peak load of 1800 RT, it will also result in very poor efficiency at night during part-load operation. A chiller combination that provides a smaller capacity chiller to operate at night and higher capacity chillers to operate during the daytime would lead to better efficiency due to better matching of chiller capacity to load.

Since performance data for chillers can normally be obtained from chiller manufacturers, if the building cooling load profile and operating hours are known, the expected energy consumption of a chiller can be computed.

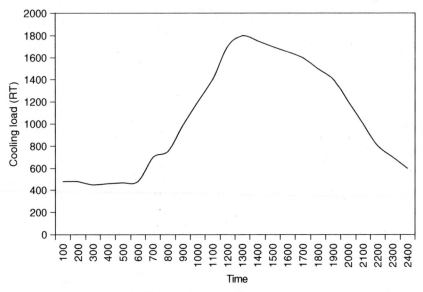

Figure 2.14 Cooling load profile for a day in a building.

Consider the part-load performance data for an 1800-RT chiller given in Table 2.8 (assumed for illustration purposes).

For the cooling load profile given in Fig. 2.14, the daily chiller energy consumption can be estimated, as shown in Table 2.9, using the following equation,

$$\text{kWh consumption} = \text{cooling load in RT} \times \text{hours of operation}$$

$$\times \text{ efficiency (kW/RT) of chiller.}$$

Similarly, the energy consumption for two other combinations using chillers of 1000-, 650-, and 500-RT capacity (part-load data given in Table 2.10) to meet the same building cooling load can be computed as shown in Tables 2.11 and 2.12.

The energy consumption for the three different chiller combinations is summarized in Table 2.13.

TABLE 2.8   Part-Load Performance Data for a 1800-RT Capacity Chiller

Percent loading	Cooling capacity (RT)	Efficiency (kW/RT)
100	1800	0.58
90	1620	0.57
80	1440	0.57
70	1260	0.57
60	1080	0.58
50	900	0.60
40	720	0.64
30	540	0.71
20	360	0.91

TABLE 2.9   Estimated Hourly Energy Consumption for an 1800-RT Chiller

Time	Load (RT)	Available capacity (RT)	Loading (%)	Efficiency (kW/RT)	Hourly power consumption (kW)
0100	480	1800	27%	0.80	384
0200	480	1800	27%	0.80	384
0300	450	1800	25%	0.80	360
0400	460	1800	26%	0.80	368
0500	470	1800	26%	0.80	376
0600	480	1800	27%	0.80	384
0700	700	1800	39%	0.64	448
0800	750	1800	42%	0.64	480
0900	984	1800	55%	0.59	581
1000	1200	1800	67%	0.57	689
1100	1400	1800	78%	0.57	798
1200	1700	1800	94%	0.58	978
1300	1800	1800	100%	0.58	1044
1400	1750	1800	97%	0.58	1015
1500	1700	1800	94%	0.58	978
1600	1650	1800	92%	0.57	941
1700	1600	1800	89%	0.57	912
1800	1500	1800	83%	0.57	855
1900	1400	1800	78%	0.57	798
2000	1200	1800	67%	0.57	689
2100	1000	1800	56%	0.59	590
2200	800	1800	44%	0.62	496
2300	700	1800	39%	0.64	448
2400	600	1800	33%	0.69	414
				Total	15,408 kWh/day

The above example shows that the energy consumption for the different options vary from one to another and that sizing and configuration of chillers has a significant effect on the cooling-energy consumption. This emphasizes the importance of accurately estimating the expected

TABLE 2.10   Part-Load Performance Data for 500-, 650-, and 1000-RT Capacity Chillers.

Percent loading	Efficiency (kW/RT) 500-RT chiller	Efficiency (kW/RT) 650-RT chiller	Efficiency (kW/RT) 1000-RT chiller
100	0.60	0.59	0.59
90	0.60	0.59	0.58
80	0.60	0.59	0.58
70	0.61	0.60	0.59
60	0.62	0.61	0.60
50	0.65	0.63	0.63
40	0.71	0.68	0.68
30	0.79	0.75	0.75
20	0.98	0.95	0.90

TABLE 2.11   Estimated Energy Consumption for Chiller Combination of Two 1000-RT Chillers

Time	Load (RT)	Available capacity (RT)	Loading (%)	Efficiency (kW/RT)	Hourly power consumption (kW)
0100	480	1000	48%	0.64	307
0200	480	1000	48%	0.64	307
0300	450	1000	45%	0.66	297
0400	460	1000	46%	0.65	299
0500	470	1000	47%	0.65	306
0600	480	1000	48%	0.64	307
0700	700	1000	70%	0.59	413
0800	750	1000	75%	0.59	443
0900	984	1000	98%	0.59	581
1000	1200	2000	60%	0.60	720
1100	1400	2000	70%	0.59	826
1200	1700	2000	85%	0.58	986
1300	1800	2000	90%	0.58	1044
1400	1750	2000	88%	0.58	1015
1500	1700	2000	85%	0.58	978
1600	1650	2000	83%	0.58	957
1700	1600	2000	80%	0.58	928
1800	1500	2000	75%	0.59	885
1900	1400	2000	70%	0.59	826
2000	1200	2000	60%	0.60	720
2100	1000	1000	100%	0.59	590
2200	800	1000	80%	0.58	464
2300	700	1000	70%	0.59	413
2400	600	1000	60%	0.60	360
				Total	14,971 kWh/day

cooling load profile for a building and then selecting the most suitable combination of chillers. When selecting the best combination of chillers, one should also keep in mind that although one combination may have a lower energy consumption, it may have a higher associated capital cost. Therefore, it is advisable to use a life-cycle cost comparison to identify the most economical solution.

The next example illustrates how chiller sizing and configuration can be used to improve the efficiency of chillers in an existing building.

The improvement in chiller efficiency possible due to correct sizing and configuration of chillers is the same whether it is a new building installation or an existing one. However, better results can normally be achieved when retrofitting an existing chiller system since the actual building cooling load profile can be measured accurately as opposed to having to use simulation tools to predict the cooling load profile for a new building.

TABLE 2.12   Estimated Energy Consumption for the Combination of Two 650-RT and one 500-RT Chillers

Time	Load (RT)	Available capacity (RT)	Loading (%)	Average efficiency (kW/RT)	Hourly power consumption (kW)
0100	480	500	96%	0.60	288
0200	480	500	96%	0.60	288
0300	450	500	90%	0.60	270
0400	460	500	92%	0.60	276
0500	470	500	94%	0.60	282
0600	480	500	96%	0.60	288
0700	700	1150	61%	0.62	434
0800	750	1150	65%	0.61	458
0900	984	1150	86%	0.60	590
1000	1200	1300	92%	0.59	708
1100	1400	1800	78%	0.60	840
1200	1700	1800	94%	0.60	1020
1300	1800	1800	100%	0.60	1080
1400	1750	1800	97%	0.59	1033
1500	1700	1800	94%	0.60	1020
1600	1650	1800	92%	0.59	974
1700	1600	1800	89%	0.59	944
1800	1500	1800	83%	0.60	900
1900	1400	1800	78%	0.60	840
2000	1200	1300	92%	0.59	708
2100	1000	1150	87%	0.60	600
2200	800	1150	70%	0.61	488
2300	700	1150	61%	0.62	434
2400	600	650	92%	0.59	354
				Total	15,116 kWh/day

TABLE 2.13   Annual Energy Consumption for the Three Different Chiller Combinations

Chiller combination	Daily kWh consumption	Annual kWh consumption
1 × 1800 RT	15,408	5,623,920
2 × 1000 RT	14,971	5,464,415
1 × 500 RT + 2 × 650 RT	15,116	5,517,340

Example 2.5 (retrofit)   Consider a building that operates daily from 7 a.m. to 10 p.m., where two chillers are operated from 11 a.m. to 9 p.m. and only one chiller at other times. The capacity of the chillers is rated at 550 RT each.

The building cooling load profile is given in Fig. 2.15, while the chiller part-load efficiency data for the existing chillers are given in Table 2.14.

Since the chillers normally operate at 375 RT each (750-RT combined load shared by two chillers) during most of the day, the resulting operating efficiency is about 0.8 kW/RT, which is a drop of 14 percent compared to the full-load effi-

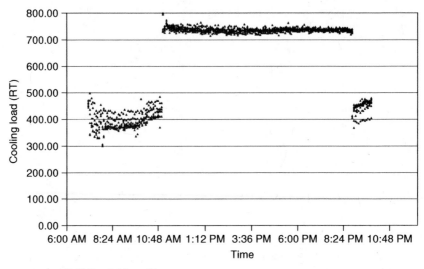

Figure 2.15 Building load profile.

ciency of the chiller. This indicates that the chillers are significantly oversized for the application. Since the chillers are rated at 550 RT, they each operate at less than 70 percent of their original capacity at the peak load of 750 RT.

This chiller system can be retrofitted to improve the efficiency by replacing the chillers with new machines sized to match the building's cooling load profile. Based on the measured peak load of 750 RT for the building, the maximum chiller operating capacity required during the warmer periods of the year is estimated to be not more than 900 RT (extra 20 percent capacity). Since the off-peak cooling load is about 450 RT for the building, a suitable capacity for the new chillers would be 450 RT each, where two machines can be operated to meet peak load while only one is operated at other times. This will ensure that the chillers operate at their most efficient operating range of 80 to 100 percent of their rated capacity.

Using the part-load efficiency of the proposed new chillers (typical data given in Table 2.15), the expected savings can be estimated as shown in Table 2.16.

TABLE 2.14 Part-Load Efficiency Data for the Existing Chillers

Chiller load (RT)	Loading (%)	Efficiency (kW/RT)
550	100%	0.70
495	90%	0.72
440	80%	0.78
385	70%	0.80
330	60%	0.85
275	50%	0.88
220	40%	0.90

TABLE 2.15   Typical Part-Load Data for the New Chillers

Chiller load (RT)	Loading (%)	Efficiency (kW/RT)
450	100%	0.54
405	90%	0.55
360	80%	0.57
315	70%	0.59
270	60%	0.61
225	50%	0.65
180	40%	0.69

TABLE 2.16   Estimated Energy Savings with New Chiller

Daily operating hours	Hours/ day	Cooling load (RT)	Present chiller efficiency (kW/RT)	Present consumption (kWh/day)	Proposed chiller efficiency (kW/RT)	Proposed consumption (kWh/day)
0700–0800	1	375	0.8	300	0.56	210
0800–0900	1	375	0.8	300	0.56	210
0900–1000	1	375	0.8	300	0.56	210
1000–1100	1	375	0.8	300	0.56	210
1100–1200	1	735	0.8	588	0.57	419
1200–1300	1	735	0.8	588	0.57	419
1300–1400	1	735	0.8	588	0.57	419
1400–1500	1	735	0.8	588	0.57	419
1500–1600	1	735	0.8	588	0.57	419
1600–1700	1	735	0.8	588	0.57	419
1700–1800	1	735	0.8	588	0.57	419
1800–1900	1	735	0.8	588	0.57	419
1900–2000	1	735	0.8	588	0.57	419
2000–2100	1	735	0.8	588	0.57	419
2100–2200	1	440	0.77	339	0.54	238
Total	15			7419		5268

$$kWh\ savings = (7419 - 5268)\ kWh/day$$
$$= 2151 \times 365\ kWh/year$$
$$= 785{,}115\ kWh/year.$$

If the average electricity tariff is $0.10/kWh, kWh savings

$$= 785{,}115\ kWh/year \times \$0.10\ /kWh$$
$$= \$78{,}512\ per\ year$$

kW peak demand savings, if applicable, is calculated as follows:

Present maximum power demand at peak cooling load

$$= 735\ RT \times 0.8\ kW/RT = 588\ kW$$

Maximum power demand with new chillers

$$= 735\ RT \times 0.57\ kW/RT = 419\ kW$$

Peak demand savings

$= (588 - 419)$ kW/month

$= 169$ kW/month

If peak demand is charged monthly at \$10/kW, Peak demand savings

$= 169$ kW/month $\times$ \$10 $\times$ 12 months

$= \$20,280$/year

Therefore, total annual savings

$= \$78,512 + \$20,280$

$= \$98,792$

2.3.4 Consolidation of chiller plant

Some facilities have more than one chiller plant or have stand-alone direct-expansion (DX) systems serving some areas while a central chiller plant serves the other areas. Such stand-alone systems and multiple chiller plants tend to be less efficient than a single central chiller plant. Therefore, in facilities that have more than one chiller plant or those with stand-alone systems serving specific areas of a building, they can be consolidated into one central chiller plant to improve the overall energy efficiency of the cooling system.

One of the main reasons for better energy efficiency of central cooling plants is the use of higher capacity chillers as compared to the use of small chillers in the case of multiple plants. Higher capacity chillers used in central plants usually have a higher rated efficiency than smaller chillers and therefore result in better operating efficiency. Further, for central plants, load diversity and the availability of multiple chillers leads to better matching of chiller capacity to cooling load, resulting in increased efficiency.

Applications for such consolidation of plants include building complexes that have an office tower and a retail podium served by two different chiller plants, and buildings, such as hotels, that have a number of blocks, each served by its own chiller plant. The savings that can be achieved by such chiller plant consolidation is illustrated in Example 2.6.

Example 2.6 Consider the following example of a building facility having one chiller plant serving the office tower while another chiller plant serves the podium retail block.

The chiller plant serving the office block has three 500-RT capacity chillers and operates from 8 a.m. to 10 p.m. Out of the three chillers; only two are operated at peak load, leaving the third machine on standby.

As shown in Table 2.17, the load for the office tower ranges between 500 and 650 RT during the daytime and drops to 150 RT in the evening. This kind of cooling load profile, where the evening cooling load is very much less than the daytime load, is common and is due to reduced heat gain through the building envelope and reduced occupancy during late evenings. The chiller loading (cooling load divided by the operating chiller capacity) varies from 55 to 100 percent during daytime to 30 percent during late evenings. The low chiller loading in the

TABLE 2.17   Energy Consumption for an Office Tower Chiller Plant

Hours of of operation	Number of hours A	Cooling load (RT) B	Chiller loading	Chiller efficiency (kW/RT) C	Present kWh consumption A × B × C
0800–0900	1	500	100%	0.57	285
0900–1000	1	500	100%	0.57	285
1000–1100	1	550	55%	0.74	407
1100–1200	1	600	60%	0.7	420
1200–1300	1	650	65%	0.72	468
1300–1400	1	600	60%	0.7	420
1400–1500	1	600	60%	0.7	420
1500–1600	1	550	55%	0.74	407
1600–1700	1	500	100%	0.57	285
1700–1800	1	500	100%	0.57	285
1800–1900	1	350	70%	0.68	238
1900–2000	1	150	30%	0.95	142.5
2000–2100	1	150	30%	0.95	142.5
2100–2200	1	150	30%	0.95	142.5
				Total	4347.5

evenings is because a 500-RT capacity chiller has to be operated to satisfy a load of 150 RT. As a result, the chiller efficiency drops significantly during this period.

Similarly, the chiller plant serving the retail podium block has two 400-RT capacity chillers and operate from 10 a.m. to 10 p.m. Out of the two chillers, only one is operated, leaving the second machine on standby.

As shown in the Table 2.18, the load for the podium block ranges between 300 and 350 RT during the daytime and drops to 200 RT in the evening. The chiller loading varies from 75 to 88 percent during daytime and 50 percent during late evenings. The low chiller loading in the evenings is because a 400-RT capacity chiller has to

TABLE 2.18   Energy Consumption for a Chiller Plant Serving a Retail Podium

Hours of operation	Number of hours A	Cooling load (RT) B	Chiller loading	Chiller efficiency (kW/RT) C	Present kWh consumption A × B × C
1000–1100	1	300	75%	0.7	210
1100–1200	1	325	81%	0.7	228
1200–1300	1	350	88%	0.68	238
1300–1400	1	350	88%	0.68	238
1400–1500	1	325	81%	0.7	228
1500–1600	1	325	81%	0.7	228
1600–1700	1	325	81%	0.7	228
1700–1800	1	300	75%	0.7	210
1800–1900	1	300	75%	0.7	210
1900–2000	1	300	75%	0.7	210
2000–2100	1	200	50%	0.95	190
2100–2200	1	200	50%	0.95	190
				Total	2606

be operated to satisfy a load of 200 RT. As a result, the chiller efficiency drops significantly during this period (chiller efficiencies used are based on typical data).

To satisfy the requirements of both the office tower and the retail podium, the central plant can be a new plant with new chillers selected to match the combined cooling load or one of the existing plants converted to serve as the central plant. If an existing chiller plant is converted to the central plant, the existing chillers can be used with or without additional chillers, depending on the combined cooling load. Care should be taken to ensure that all areas of the new combined system can withstand the higher static pressure imposed by the office tower chilled water distribution system. If required, heat exchangers can be installed to hydraulically isolate the office tower and retail podium systems.

The most logical choice in this case would be to use the chiller plant serving the office tower to act as the combined central plant. The analysis for this scenario is given in Table 2.19.

Since this plant has three 500-RT chillers, only two machines need to be operated during daytime to meet the expected combined cooling load. Due to the load diversity of the two blocks, the chillers would be better loaded during most times of the day, leading to better chiller efficiency. The best improvement in efficiency would be achieved in the late evenings when only one 500-RT chiller has to be operated to satisfy the combined load of 350 RT as compared to the need for operating two chillers previously.

The savings in this case is the difference between the kWh consumption for the two separate plants and the central combined plant.

Therefore, savings

$$= [(4347.5 + 2606) - 5772]$$

$$= 1181.5 \text{ kWh/day}$$

$$= 431,248 \text{ kWh/year.}$$

TABLE 2.19 Energy Consumption for Proposed Combined Chiller Plant

Hours of operation	Number of hours A	Combined cooling load (RT) B	Average chiller loading	Average chiller efficiency (kW/RT) C	Proposed kWh consumption A × B × C
0800–0900	1	500	100%	0.57	285
0900–1000	1	500	100%	0.57	285
1000–1100	1	850	85%	0.55	468
1100–1200	1	925	93%	0.56	518
1200–1300	1	1000	100%	0.57	570
1300–1400	1	950	95%	0.57	542
1400–1500	1	925	93%	0.56	518
1500–1600	1	875	88%	0.55	481
1600–1700	1	825	83%	0.55	454
1700–1800	1	800	80%	0.57	456
1800–1900	1	650	65%	0.72	468
1900–2000	1	450	90%	0.56	252
2000–2100	1	350	70%	0.68	238
2100–2200	1	350	70%	0.68	238
				Total	5772

In this particular example, the cost that needs to be incurred to achieve this saving would be to modify the chilled water piping and pumping systems to enable the office tower plant to serve the podium block. This makes it financially attractive when compared to constructing a new central plant with new chillers.

The decision whether to use one of the existing plants with or without the existing chillers as compared to constructing a new plant depends on factors such as whether the existing chillers can satisfy the combined load, or if additional chillers are required, whether they can be installed in the existing plant room, and the efficiency of the existing chillers. Usually, if the chillers are old and inefficient, it would be better to replace them with new machines suitably sized to satisfy the combined load.

2.3.5   Chiller sequencing

As explained earlier, the operating efficiency of chillers depends on their loading. A typical chiller loading versus efficiency relationship is shown in Fig. 2.13. As can be seen from this figure, the chiller efficiency is best when operating in the range 60 to 100 percent of its capacity, while optimum efficiency is at 80 percent loading (some chillers operate best at 100 percent).

It was also explained earlier that it is important to have the right combination of chillers so that the operating chiller capacity can be matched to the varying building cooling load to ensure that the chillers are always able to operate within the best efficiency range. However, to achieve this objective, it is not sufficient to just have the correct combination of chillers. It is also necessary to have a control system to ensure that the matching of chiller capacity to load can be done continuously.

This is normally achieved through chiller sequencing, which enables operation of the most efficient combination of chillers to meet the varying building load. A chiller sequencing programme will start an additional chiller when the building load exceeds the capacity of the running chillers and stop a chiller when the remaining chillers can handle the building load.

If chiller operations are optimized based only on chiller efficiency, the optimal point for adding or removing chillers may occur when the chillers are not fully loaded since maximum chiller efficiency usually occurs at part load (Fig. 2.13). For example, theoretically it may be better to operate two chillers at part load rather than one big chiller at 100 percent capacity. However, this strategy ignores the need for additional chilled water pumps, condenser water pumps, and cooling towers when the additional chiller is operated. Therefore, optimizing should be carried out based on the total power consumption of the chiller system (chillers, pumps, and cooling towers), as shown in Fig. 2.16.

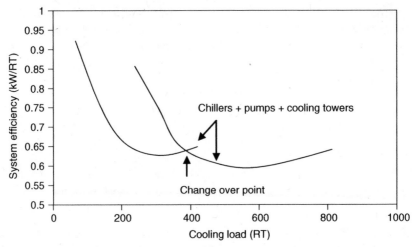

Figure 2.16 Typical system efficiency curves for different chiller operating scenarios.

Normally for systems where chillers have dedicated pumps and cooling towers, the operating chillers should be fully loaded before starting an additional chiller because the power required for operating the pumps and cooling towers for the additional chiller is greater than the saving in chiller power when operating it at optimum part-load conditions.

However, for systems that do not have dedicated pumps (systems with variable speed primary pumps or multiple primary pumps on a common header), the optimal load condition for bringing chillers online or offline may not occur at chiller full load. In such a case, the optimal chiller switching point will depend on the combined chiller, pumping, and cooling tower power consumption.

In most buildings, the relationship between cooling load and the chilled water flow required to meet that load are not linear. Therefore, often chillers need to be sequenced to satisfy both building cooling load and chilled water flow requirements. The actual relationship between chilled water flow and the capacity of typical constant air volume (CAV) AHU cooling coils is shown in Fig. 2.17. As the figure shows, the relationship between flow and capacity is logarithmic and also changes with chilled water temperature. Further, other system operating conditions, such as unequal loading of AHUs or low ΔT (low temperature difference between chilled water return and supply), can cause the relationship between flow and capacity in a chilled water system to be non-linear. Figure 2.18 shows the ideal linear relationship with possible extremes for the chilled water flow and capacity relationships for typical chilled water systems.

As Fig. 2.18 shows, generally for the lower curve, a higher percentage of flow is required to satisfy a particular load (e.g. 80 percent flow

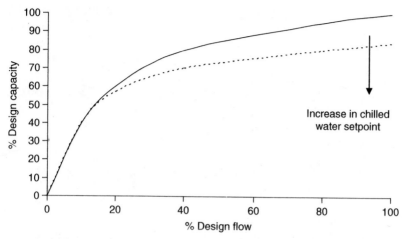

Figure 2.17 Typical cooling coil characteristic for a constant air volume AHU

to satisfy 50 percent capacity). Therefore, if this particular installation has two equal sized chillers, each sized to meet 50 percent of the total load, the second chiller has to be switched on when the building load exceeds about 25 percent of the total because at this point, more than 50 percent flow (maximum flow for one set of chiller and pump) is required to satisfy the load. In such situations, an additional chilled water pump (and therefore an additional chiller) needs to be turned on to satisfy flow requirements although the operating chiller has not been fully loaded.

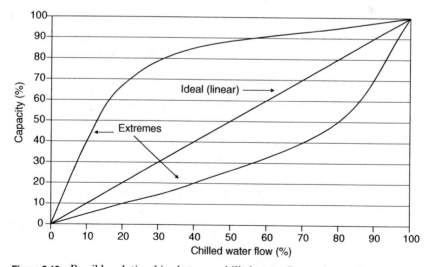

Figure 2.18 Possible relationships between chilled water flow and capacity.

Another important operating parameter to be considered in chiller sequencing is chiller motor loading. The motor of the chiller compressor is sized to meet the requirements of the compressor when it is fully loaded. However, the actual current drawn by the compressor motor at a particular cooling load also depends on the chilled water and condenser water temperatures. If for instance, the chilled water supply temperature is set higher during off-peak cooling periods or if the condenser water supply temperature is lower due to favorable weather conditions, the current drawn by the motor at a particular load will be less since the compressor has to work against a lower pressure differential (explained later). Therefore, when a chiller reaches its maximum design capacity, the motor may not be loaded to 100 percent of the design value and as a result the chiller can be loaded further until the motor reaches its maximum capacity, resulting in the chiller providing a higher capacity than rated. If, in such an instance, the chillers are sequenced only based on cooling load, an additional chiller will be turned on before the running chiller is loaded fully.

Therefore, it can be seen that it is not sufficient just to sequence chillers based on the cooling load since other parameters such as chilled water flow requirements and motor loading need to be considered. A few typical chiller sequencing strategies of different complexities are illustrated next.

Typical chiller sequencing strategies

a. Simple control strategy using chilled water temperature to sequence chillers. In the system shown in Fig. 2.19, an additional chiller is turned on, when the temperature of chilled water leaving the chillers T_1, T_2, or T_3 is greater than the set point (chiller/s in operation cannot satisfy load).

Similarly, if two chillers are in operation and they are of equal capacity, an operating chiller is switched-off if chiller ΔT/design ΔT

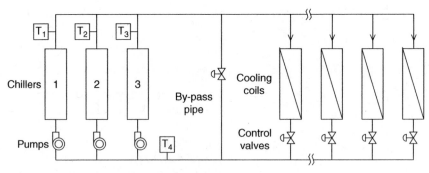

Figure 2.19   Chiller sequencing using chilled water temperature.

is less than 0.5 (cooling load is less than half and can be satisfied by one chiller) and chilled water leaving temperature is not greater than set point. The chiller ΔT and design ΔT are the temperature differences between chilled water return and chilled water supply under operating conditions and design conditions, respectively.

This is a simple but crude way of sequencing chillers. Since parameters such as chilled water flow, cooling load and motor loading are not measured, chiller operations cannot be optimized.

b. Commonly applied control strategy using cooling load and chilled water temperature (Fig. 2.20).

In this system, the cooling load is calculated using the temperature difference for chilled water $(T_2 - T_1)$ and the chilled water flow rate. An additional chiller is turned-on when the temperature of chilled water leaving the chiller T_1 is greater than set point (chiller/s in operation cannot satisfy load) or if the cooling load is equal to the capacity of operating chillers.

An operating chiller is switched off if the cooling load is less than the capacity of operating chillers, less the capacity of one chiller, and T_1 is not greater than set point.

This is a better way of sequencing chillers since chillers are operated based on the cooling load. However, as pointed out earlier, chillers can provide more than the design capacity in some off-design operating conditions, such as lower condenser water supply temperature or higher chilled water temperature. Therefore, in such an instance, an additional chiller will be switched-on before the operating chillers are fully loaded. The next strategy is designed to overcome this shortcoming.

c. Recommended strategy for chiller sequencing in systems with only primary pumping. This strategy uses cooling load, chilled water temperature, and chiller motor loading to sequence chillers (Fig. 2.21).

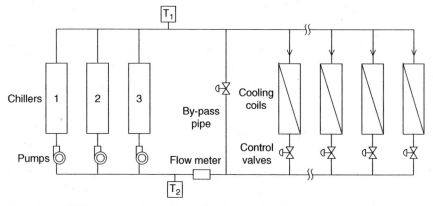

Figure 2.20 Chiller sequencing using cooling load and chilled water temperature.

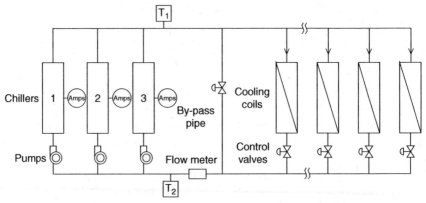

Figure 2.21   Chiller sequencing using cooling load, chilled water temperature and motor loading.

An additional chiller is turned on when the temperature of chilled water leaving the chiller T_1 is greater than set point (chiller/s in operation cannot satisfy load) or if the cooling load is equal to the capacity of the operating chiller and the current drawn by the chiller motor is equal or greater than the rated full load current of the motor (fully loaded motor).

An operating chiller will be switched off if the cooling load is less than the capacity of operating chillers, less the capacity of one chiller, and T_1 is not greater than set point.

This method of sequencing chillers takes into consideration the loading of chiller motors, which helps ensure that chillers are fully loaded at off-design operating conditions, such as at lower condenser water supply temperature and higher chilled water temperature, when chillers are able to provide higher than the rated capacity.

d. Simple sequencing strategy for primary-secondary pumping systems. Primary-secondary pumping systems consist of two chilled water pumping loops, a primary loop, which pumps chilled water through the chillers, and a secondary loop, which pumps chilled water to the terminal units such as AHUs and FCUs. These two loops are hydraulically decoupled from each other (explained in detail in Chapter 4).

As shown in Fig. 2.22, the simplest strategy is to use a bidirectional flow meter on the decoupler pipe to indicate whether the chilled water flow in the decoupler pipe is from the supply side to the return side or in reverse, from the return side to supply side. If the flow is from the return side to the supply side, it indicates that the chilled water flow in the secondary loop is higher than that in the primary loop and therefore an additional primary pump (and a chiller) needs

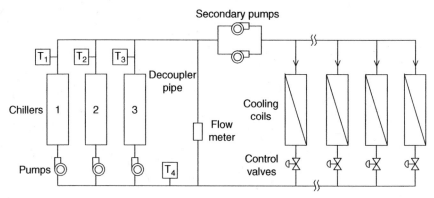

Figure 2.22 Chiller sequencing for primary-secondary systems using decoupler flow and chilled water temperature.

to be turned-on. Similarly, if the temperature of the chilled water leaving the chillers (T_1, T_2, or T_3) is greater than set point, it indicates that the chillers in operation are unable to meet the cooling load and an additional chiller needs to be operated.

If the flow in the decoupler is from supply side to return side and the flow is greater than 110 percent of the flow of one chiller, it indicates that if one chiller is switched off, the chilled water flow requirements can still be met. Simultaneously, the temperature difference between return and supply chilled water is compared with the design ΔT (like in the earlier example for primary only systems) to ensure that if one chiller is switched off, the resulting ΔT will not exceed the design value. For example, when two equal capacity chillers are in operation, if the value of the chilled water ΔT/design ΔT is less than 0.5, the cooling load can be satisfied by only one chiller.

One of the shortcomings of this system is that it does not sequence chillers based on actual cooling load or loading of chillers, but is based on indirect measurements. A better control strategy is described next.

e. Recommended strategy for chiller sequencing in systems with primary-secondary pumping. It is recommended that two flow meters be used to measure the chilled water flow; one on the primary chilled water loop and the other on the secondary chilled water loop, as shown in Fig. 2.23. The difference in readings between the two flow meters will yield the flow rate and direction of chilled water flow in the decoupler pipe.

This indirect measurement of decoupler flow is more accurate than direct measurement using a flowmeter on the decoupler pipe, where the flow can vary from zero to more than the design flow for one chiller.

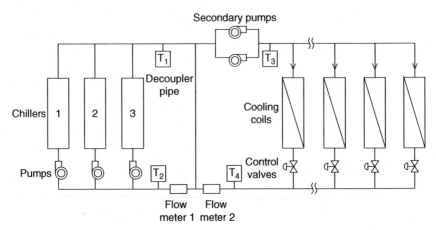

Figure 2.23 Chiller sequencing for primary-secondary systems using individual flow meters and chilled water temperatures.

In this system, an additional chiller is switched on if the chilled water supply temperature (T_1) is greater than set point (chillers cannot satisfy cooling load) or if the chilled water flow in the primary loop is less than the flow in the secondary loop (insufficient chilled water flow).

Similarly, an operating chiller can be switched off if the flow in the decoupler from the supply side to the return side is more than 110 percent of the design flow rate of one chiller and the cooling load (measured using flowmeter and chilled water temperature difference) is less than the total capacity of the operating chillers less the capacity of one.

2.3.6 Reset of chilled water temperature

Figure 2.24 shows the p-h diagram for an ideal vapor compression cycle. "Work" is done in the cycle by the compressor to compress the refrigerant vapor from the evaporation pressure to the condensing pressure. Therefore, if the evaporating pressure is increased or the condensing pressure is reduced (explained later), the amount of work to be done by the compressor reduces. This will result in an increase in COP or reduction in the kW/RT of the chiller. In general, it is estimated that improvement in chiller efficiency of 1 to 2 percent can be achieved by increasing the chilled water temperature by 0.6°C. Figure 2.25 shows the improvement in efficiency due to chilled water reset for a typical chiller.

Chiller systems are generally designed for chilled water supply at 6.7°C (44°F) to meet the design peak cooling load. However, chillers seldom have to operate under full-load conditions. Under such operating

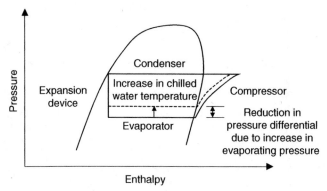

Figure 2.24 Pressure–enthalpy diagram showing effect of increasing chilled water temperature.

conditions, most of the time it would not be necessary to provide chilled water at the design value and therefore the chilled water temperature can be reset upwards.

One method of resetting the chilled water temperature is by monitoring the position of the control valves on the cooling coils. The valve that is open the most can be used to reset the temperature and the chilled water temperature can be reset upwards in steps until this control valve or any other reaches a preset value (eg. maximum 90 percent open). Similarly, the chilled water temperature can be reset downwards if any control valve opens beyond this set maximum value.

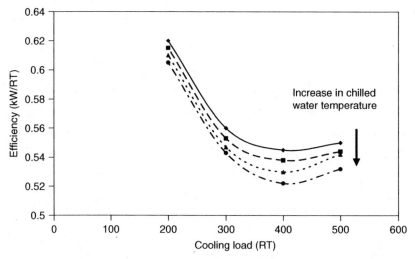

Figure 2.25 Typical relationship between chiller efficiency and chilled water temperature.

Another method used for resetting the chilled water temperature is based on the return chilled water temperature. When the return chilled water temperature reduces (eg. from 12 to 10°C), the chilled water supply temperature can be raised to bring the return temperature back to the design value (12°C). This works on the assumption that the return chilled water temperature drops if the cooling load drops, which will be the case only if the chilled water flow is constant. Therefore, in variable flow systems, where the flow reduces when the load drops, the return temperature will not indicate load. Further, even if the system is a constant flow one, if the loads are not homogeneous, the return temperature will be the average temperature of water returning from the different loads and will only indicate the average load. This could lead to situations wherein the chilled water temperature will be reset upwards (because the mixed chilled water return temperature is low) when some areas experience full load conditions and need chilled water at the design temperature to satisfy the load.

The chilled water temperature can also be reset upwards based on the cooling load or outdoor temperature. Since reduction in outdoor temperature leads to lower cooling load, the outdoor temperature or a direct measurement of the load can be used to reset the chilled water temperature.

However, when the chilled water supply temperature is increased, the chilled water flow would have to be increased to satisfy the same load for CAV systems due to the performance characteristic of the coils (Fig. 2.17).

In variable flow chilled water pumping systems, when the chilled water temperature is raised to improve chiller efficiency, the power consumed by the chilled water pumps will increase due to the need for more water flow to satisfy the same load. Therefore, when optimizing savings from chilled water reset, the combined chiller and pumping power should be considered to ensure that savings from the chiller exceeds the extra energy consumption for chilled water pumping.

As Fig. 2.26 shows, chiller efficiency improves when chilled water temperature is increased but results in a drop in chilled water pumping

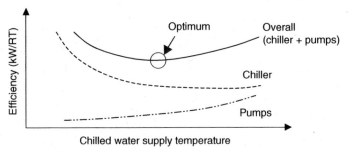

Figure 2.26 Effect of chilled water temperature on chiller and pump efficiency.

efficiency (more chilled water needs to be pumped to satisfy the same load). The optimum operating point is when the overall efficiency (chiller and pumps) is the highest. This optimum operating point will vary for different installations as it depends on chiller and pump performance characteristics.

Further, in variable air volume (VAV) air distribution systems, increase in chilled water temperature leads to the need for higher supply airflow from the AHUs to satisfy the cooling load. This too results in higher AHU fan power consumption.

Based on research, at common AHU supply air temperatures of 11 to 13°C, the chilled water set point can be increased to about 8.5°C (47°F), while maintaining the total system (chillers, pumps, and AHU fans) power consumption within ±1 percent of the optimum.

When resetting chilled water temperature, one should also bear in mind that cooling coils of air-conditioning systems not only provide sensible cooling but also remove moisture from the air in the air-conditioned space and help to maintain the relative humidity. The moisture removal ability of cooling coils depends on the chilled water supply temperature since the air has to be cooled to its dew point temperature to condense the moisture from the air. Therefore, raising the chilled water supply temperature could lead to a reduction in the moisture removal ability of the coils and higher relative humidity in conditioned spaces.

2.3.7 Reset of condenser water temperature

Similar to chilled water reset, the operating efficiency of chillers can also be improved by reducing the condenser water temperature. The improvement in chiller efficiency is due to the reduced pressure differential across which the compressor has to work when the condenser water temperature is reduced (Fig 2.27). The savings from condenser

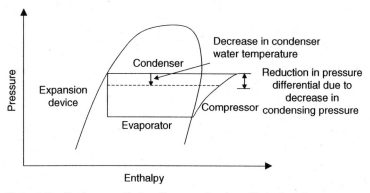

Figure 2.27 Pressure–enthalpy diagram showing effect of condenser water temperature.

water reset is similar to that for chilled water reset, and improvement in chiller efficiency of 1 to 2 percent can be achieved by reducing the condenser water temperature by 0.6°C (1°F). Figure 2.28 shows the improvement in efficiency possible due to condenser water reset for a typical chiller.

Cooling towers are designed to cool condenser water to within a few degrees of the wet-bulb temperature. This temperature difference between condenser water supply temperature and wet-bulb temperature is called the *cooling tower approach*. Therefore, when the wet-bulb temperature of the outdoor air drops, cooling towers are able to provide condenser water at a lower temperature, which helps to improve chiller efficiency.

2.3.8 Maintaining surfaces of condenser tubes

The evaporator and condenser tubes of chillers provide the surface for heat transfer between the refrigerant and chilled water or condenser water, respectively. If there is scaling or fouling on the tube surfaces, the resistance to heat transfer increases. This results in a higher temperature difference driving the heat transfer process, which leads to lower chiller efficiency.

Condenser tubes are more prone to fouling since the circulating water is open to the outside (at the cooling towers) and therefore need to be cleaned regularly. Research has shown that 0.6 mm of scale will increase the chiller compressor power consumption by 20 percent.

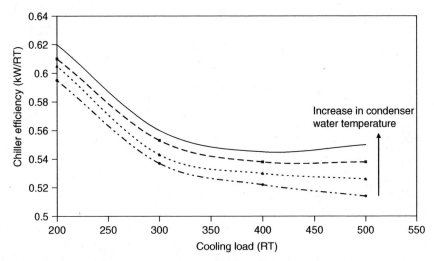

Figure 2.28 Typical relationship between chiller efficiency and condenser water temperature.

Scale and fouling in condensers is normally controlled by:

a. Water treatment systems to inhibit scaling and fouling

b. Blow down, where water is regularly drained from the bottom of the cooling tower basin and topped up with clean water

c. Tube cleaning, where the insides of tubes are cleaned periodically by brushing

The effectiveness of water treatment systems in controlling scaling depends on the particular treatment programme employed. In chemical treatment programs the alkalinity of the water, which increases the tendency for scaling, is controlled by increasing its acidity. However, increase in acidity leads to increased corrosion and, therefore, chemical treatment programs are used more for controlling scaling than completely eliminating it.

Blow down and tube cleaning too are measures that are carried out periodically, but cannot completely eliminate scaling and fouling. This can result in chiller efficiency varying with time, as shown in Fig. 2.29. When scale and fouling deposits collect on condenser tubes, the resistance to heat transfer across the tubes from the condensing refrigerant to the cooling water increases. This leads to an increase in the condensing temperature (and pressure) and causes the chiller efficiency to drop since the compressor has to compress the refrigerant vapor to a higher pressure. Over time, the thickness of the layer formed by scale and fouling increases, resulting in a gradual drop in chiller efficiency. Periodically, when tubes are cleaned by brushing, the efficiency returns to the normal rated value. Thereafter, the efficiency gradually drops again when scale and fouling builds up on tube surfaces. This results in a kind of "saw tooth" pattern, where the length of the tooth denotes the time interval between cleanings while the height of the tooth indicates the maximum drop in chiller efficiency between cleanings.

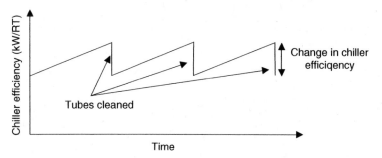

Figure 2.29 Variation of chiller efficiency due to scaling and fouling of condenser tubes.

To avoid this, automatic tube cleaning systems (with balls or brushes), which clean the condenser tubes periodically during operation of the chillers, can be used to maintain chiller efficiency.

Figure 2.30 shows a typical system using sponge balls for cleaning tubes. These sponge balls have a slightly larger diameter than the inside bore of the condenser tubes. The balls are circulated through the tubes at regular intervals. As the balls travel through the tubes, scale and fouling deposits are removed. After passing through the tubes, the balls are collected by a strainer. Thereafter, the balls are returned to the cleaning section for automatic cleaning and are injected back again after a set time.

The financial viability of installing such an automatic tube cleaning system depends on the drop in chiller efficiency before the tubes are cleaned. Since chiller efficiency cannot be easily measured and since it is even harder to detect a drop in chiller efficiency, the condenser water "approach temperature" can be used as an indicator of scale and fouling build up on condenser tubes. The condenser water approach temperature is the difference between the condensing refrigerant temperature and the temperature of the condenser water leaving the condenser. A high approach temperature usually indicates that tube cleaning is necessary. Chiller condensers are normally designed for an approach of about 1°C and therefore chillers that operate at approach temperatures higher than this would be good candidates for the installation of automatic cleaning systems. However, the actual savings that can be achieved will depend on how much the approach temperature can be reduced and the actual operating hours of the chiller.

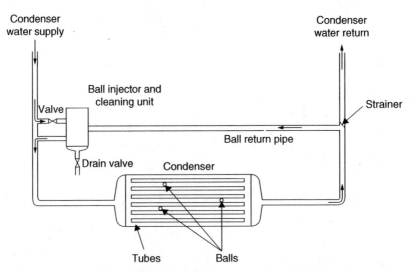

Figure 2.30 Typical arrangement of an automatic condenser tube cleaning system.

2.3.9 Dedicated chillers for night operation

The cooling load of a building is dependent on the amount of heat gain from internal and external sources. External heat gain is mainly due to conduction of heat and transmission of solar radiation through the building facade, while internal heat gain is due to heat generated by the building's occupants, lighting, and other equipment.

The cooling load varies during the day mainly due to changes in external factors such as outdoor temperature and intensity of solar radiation. In addition, changes in internal factors such as occupancy also cause the cooling load to vary.

As a result, buildings that operate 24 hours a day, like hotels or residential apartments, experience much lower cooling loads during the night as compared to the daytime. Therefore, a chiller or a combination of chillers, which have been sized to meet the daytime cooling load of a building, may be oversized for night load operation. This causes the chillers to operate at lower efficiency since chiller efficiency depends on their loading. This is illustrated further in Example 2.7.

Similar operating conditions that lead to inefficient operation of chillers may also be experienced in buildings that do not operate 24 hours a day, but operate during late evenings or weekends.

Example 2.7   Consider a case where the building's cooling load varies as shown in Fig. 2.31. The daytime cooling load varies from about 1000 to 1500 RT while the night load is about 200 to 250 RT. This low night cooling load is experienced daily from 10 p.m. to 8 a.m. During daytime, two to three chillers of 600 RT capacity

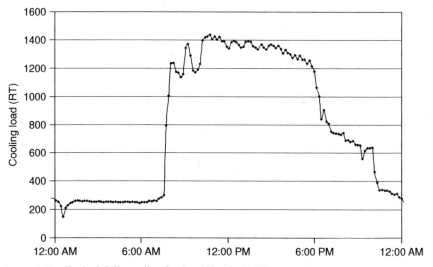

Figure 2.31   Typical daily cooling load profile of a building.

TABLE 2.20   Chiller Performance Data

Loading (%)	Chiller load (RT)	Efficiency (kW/RT)
100	600	0.64
90	540	0.63
80	480	0.63
70	420	0.65
60	360	0.68
50	300	0.73
40	240	1.0
30	180	1.2

are operated, while only one chiller of 600-RT capacity is operated at night to meet the building's cooling load.

During daytime, when two to three chillers are operated to meet the cooling load of 1000 to 1500 RT, chillers are loaded to more than 70 percent of their capacity. Chiller efficiency during daytime is therefore about 0.64 kW/RT (Table 2.20). However, at night, when a 600-RT chiller has to operate to meet a load of 200 to 250 RT, it is loaded to only about 30 to 40 percent of the capacity. This results in the chiller efficiency dropping to below 1.0 kW/RT.

The solution, in this case, is to install an additional small chiller to act as the night chiller. The new chiller capacity can be 250 RT to meet the night load of 200 to 250 RT. The efficiency of a 250-RT chiller at this operating load will be much better and will help to improve the chiller efficiency at night from 1.2 kW/RT to about 0.58 kW/RT.

The measure will result in a reduction in kWh consumption due to the improved operating efficiency of the new chiller. However, it will not normally result in the reduction of peak power demand since buildings usually experience maximum power demand during daytime. An estimation of savings is illustrated in Table 2.21.

TABLE 2.21   Estimated kWh Consumption Savings During Nighttime Chiller Plant Operation

Time	Hours/day A	Cooling load (RT) B	Present chiller efficiency (kW/RT) C	Present consumption (kWh/day) A × B × C	Proposed chiller efficiency (kW/RT) D	Proposed consumption (kWh/day) A × B × D
2200–2300	1	250	1.2	300	0.58	145
2300–2400	1	250	1.2	300	0.58	145
0000–0100	1	250	1.0	250	0.58	145
0100–0200	1	250	1.0	250	0.58	145
0200–0300	1	250	1.0	250	0.58	145
0300–0400	1	250	1.0	250	0.58	145
0400–0500	1	250	1.0	250	0.58	145
0500–0600	1	250	1.0	250	0.58	145
0600–0700	1	250	1.0	250	0.58	145
0700–0800	1	250	1.0	250	0.58	145
Total	10			2600		1450

Since the capacity of the new chiller is half the capacity of existing chillers, suitable chilled water and condenser water pumps would need to be installed to match the capacity of the new chiller. Alternatively, variable speed drives can be installed on the existing pumps to enable them to be operated at a lower speed to match the needs of the new chiller. A similar strategy can also be used to enable operation of an existing cooling tower fan at lower speed to meet the requirements of the new chiller.

$$\text{kWh savings} = (2600 - 1450) \text{ kWh/day} = 1150 \text{ kWh/day}$$

If the chiller plant operates 7 days a week

$\times$ 52 weeks a year, kWh savings

$= 1{,}150 \times 7 \times 52 \text{ kWh/year} = 418{,}600 \text{ kWh/year}$

2.3.10 Use of absorption chillers

As explained earlier, absorption chillers need a heat source to operate. Usually absorption chillers operate on low-pressure steam, hot water, or are direct fired. Although such chillers are less efficient than those working on the vapor compression cycle, if a waste heat source is available, it can be utilized to operate absorption chillers without incurring any extra cost for fuel.

Waste heat from engine jacket water, engine or process exhaust gases, and various industrial processes that generate heat are good sources of heat that can be utilized to operate absorption chillers. The waste heat sources can be either fed directly into the generator of the absorption chiller or passed through a heat exchanger to recover useful heat. A typical arrangement of an absorption chiller operating using waste heat is shown in Fig. 2.32.

It should be noted that absorption chillers need higher cooling tower capacity due to the need for higher heat dissipation, which include the heat removed in the evaporator and that added in the generator. This results in added electricity consumption for cooling tower fans and condenser water pumps when compared to conventional chillers.

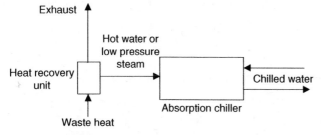

Figure 2.32 Arrangement of an absorption chiller with a heat-recovery system.

The feasibility of utilizing an absorption chiller using waste heat, instead of an electric chiller, is illustrated in Example 2.8.

Example 2.8   Consider the case of an industrial plant that has waste heat produced by a process system, which is currently released to the environment. If the plant uses a 150-RT electric chiller to provide space and process cooling, the feasibility of using an absorption chiller (using the waste heat) in place of the electric chiller can be assessed as follows:

The kW consumption for the two systems (excluding the chilled water pumps, as the consumption will be the same for both options) is estimated as follows:

	Electric chiller	Absorption chiller
Chiller	135 kW	2.5 kW
	(150 RT × 0.9 kW/RT)	
Condenser water pumps	15 kW	15 kW
Cooling tower	5.5 kW	7.5 kW
Hot water pump	—	7.5 kW
Total	155.5 kW	32.5 kW

$$\text{kWh savings} = (155.5 \text{ kW} - 32.5 \text{ kW}) \times \text{operating hours/year} \times \text{tariff}$$

$$\text{Peak demand savings (if applicable)}$$

$$= (155.5 \text{ kW} - 32.5 \text{ kW})$$

$$\times \text{ monthly demand charges} \times 12 \text{ months}$$

The following assumptions can be used to compute the simple payback period for the project:

Cost of the absorption chiller and heat exchanger = $250,000

Electricity tariff = $0.10/kWh

Peak demand charges = $10/month

Operating hours = 7000 h/year

$$\text{kWh savings} = (155.5 \text{ kW} - 32.5 \text{ kW}) \text{ kWh} \times 7000 \text{ hrs} \times \$0.10$$

$$= \$86,100 \text{ year}$$

$$\text{Peak demand savings} = (155.5 \text{ kW} - 32.5 \text{ kW}) \text{ kW} \times \$10 \times 12 \text{ months}$$

$$= \$14,760/\text{year}$$

Total savings = $86,100 + $14,760 = $100,860/year.

The simple payback period

$$= \text{cost for absorption chiller system/total savings}$$

$$= \$250,000/\$100,860$$

$$= 2.5 \text{ years.}$$

2.3.11 Thermal storage

Thermal storage is a form of storage that stores energy (cold or heat energy) for use at a later time. Common thermal storage systems enable cooling to be performed during non-peak periods and storing it for use during peak periods.

Thermal storage systems provide building owners potential for achieving cost savings if the difference in electricity tariffs between peak and off-peak is high, if peak demand charges are high, or the building maximum cooling load is much higher than the average cooling load.

The type of system depends on the storage medium. The three main storage systems used are; chilled water, ice, and eutectic salts. Chilled water storage systems (Fig. 2.33) use the sensible heat capacity of water, which is 4.18 J/g C, to store cooling capacity. Ice storage systems use the latent heat of fusion of water, which is 333 J/g, to store cooling capacity. Similarly, eutectic salts too make use of the latent heat capacity to store cooling capacity, but have the advantage of being able to operate at higher temperatures than ice systems. The storage capacity depends on the storage medium used and ice systems use the least space while chilled water systems require the most space.

Since chilled water storage systems operate at normal chiller operating conditions, standard chiller systems can be used, while for ice systems, special chillers that are designed to operate at low temperatures or ice-making systems are required.

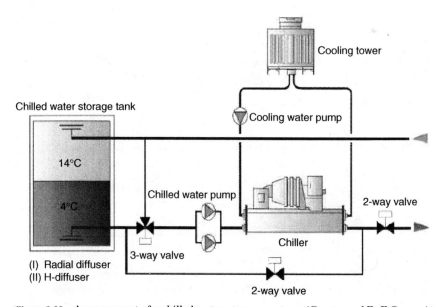

Figure 2.33 Arrangement of a chilled water storage system. (*Courtesy of EnE System.*)

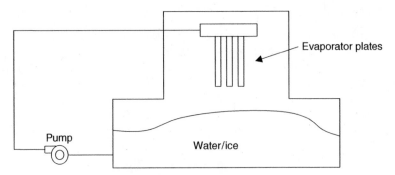

Figure 2.34 Arrangement of an ice harvesting system.

Some technologies available for producing ice are; ice harvesting systems (Fig. 2.34), ice-on-coil systems (Fig. 2.35), encapsulated ice systems (Fig. 2.36) and ice slurry systems. In ice harvesting systems, ice is formed on the surface of an evaporator and is periodically released into a storage tank partially filled with water. In the ice-on-coil systems, a coil is submerged in a tank containing water. The coolant medium, normally a glycol and water mixture from a chiller, is passed through the coil. Ice forms and accumulates on the surface of the coil. Storage is discharged by passing warm return water through the coils which melts the ice on the outside.

Encapsulated ice systems use water inside submerged plastic containers that are frozen and later thawed by passing cold coolant and

Figure 2.35 Typical ice-on-coil tanks with cutaway. (*Courtesy of Calmac.*)

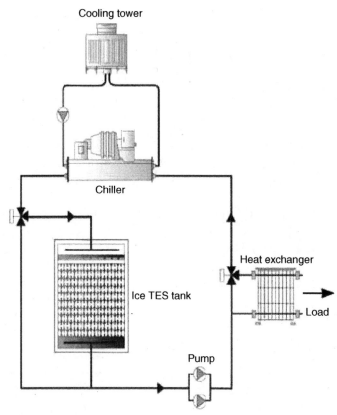

Cooling tower

Chiller

Heat exchanger

Ice TES tank

Load

Pump

Figure 2.36 Arrangement of an encapsulated ice system. (*Courtesy of EnE System.*)

warm water, respectively, outside the containers. Figure 2.36 shows a system with ball-shaped containers. Ice slurry systems store water or a mixture of water and glycol in a slurry state, which comprises of a mixture of ice crystals and liquid. The slurry is either pumped direct to the load or a heat exchanger is used to cool a secondary fluid.

Eutectic salts use a combination of inorganic salts, water, and other elements to form a mixture that freezes at a temperature above the freezing point of water. This enables the use of standard chillers for such systems.

Since in chilled water storage systems chillers operate at normal operating conditions, there is no change in chiller efficiency. However, for ice systems, since chillers have to operate at low temperatures, such as $-7°C$, there is a significant drop in chiller efficiency. The drop in efficiency for centrifugal chillers can be about 50 percent (e.g. from 0.6 to 0.9 kW/RT). However, the drop in efficiency for air-cooled chillers is not

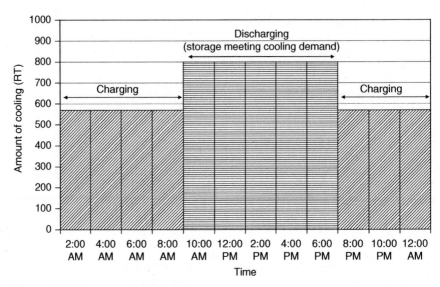

Figure 2.37 Full storage strategy.

as significant as they can make use of lower ambient temperature at night to work at a lower condenser temperature.

The two common storage strategies are full storage and partial storage (Fig. 2.37 and 2.38). In full storage systems, the entire peak cooling load is shifted to off-peak hours. The system operates at full capacity during non-peak hours to charge the storage, which is later discharged

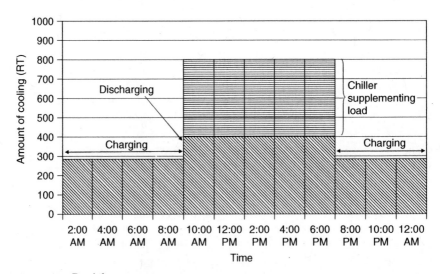

Figure 2.38 Partial storage strategy.

during the peak period. This strategy is attractive when peak demand charges are high and there is a very big difference in electricity tariff between peak and off-peak periods.

In partial storage systems, chillers operate to meet part of the cooling demand during peak periods while the remainder of the cooling demand is met by the storage system. The advantage of this system is that equipment can be sized to be less than the design maximum capacity. This strategy is attractive when the peak cooling demand is much higher than the average cooling demand.

The evaluation of whether to install a thermal storage system, and whether it should be chilled water or ice storage system can be complex and will involve the consideration of many factors. Some of the factors that need to be considered in the evaluation are; electricity tariffs, peak power demand charges, daily and seasonal cooling demand patterns, space availability, efficiency of systems, equipment cost, maintenance costs, and type and configuration of equipment. A life-cycle costing will normally be necessary for such an evaluation.

Another potential application for thermal storage systems is in buildings where a "night load" (a relatively low cooling load at night) has to be supported by high-capacity chillers sized to meet daytime cooling load, which results in very inefficient chiller operation.

In such situations, if a thermal storage system is used, it can be charged during daytime using the "day chillers" and discharged at night to satisfy the night cooling load.

2.3.12 Chiller free cooling

Chiller "free cooling" involves using a chiller to provide cooling with its compressors switched off. This strategy is possible only when outdoor weather conditions are such that the outdoor wet-bulb temperature is low enough to make condenser water colder than the chilled water and building cooling is required. Such conditions are encountered in temperate climates during autumn and spring. The chiller then operates like a heat pipe where the refrigerant evaporated by the chilled water migrates to the colder condenser, which causes the refrigerant to condense and flow back to the evaporator.

To make this system function, a pipe with a valve needs to be installed between the evaporator and condenser to enable the refrigerant to migrate from the evaporator to the condenser when the compressor is not operating. Similarly, a piping connection is necessary for the refrigerant to bypass the orifice used to control refrigerant flow to the evaporator under normal chiller operations. The arrangement of a chiller functioning in the free-cooling mode is shown in Fig. 2.39.

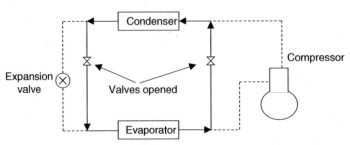

Figure 2.39   Arrangement of a chiller performing "free-cooling."

2.3.13   Waterside economizer

The waterside economizer is also a form of free-cooling that can be used in temperate climates during certain seasons to either precool or completely cool the return chilled water when the outdoor weather conditions are favorable. The strategy can be applied to water-cooled chiller systems using cooling towers.

In this system, whenever the condenser water temperature is less than the chilled water temperature, it uses cooling tower water to cool return chilled water using a heat exchanger, (Fig. 2.40). Depending on outdoor conditions and condenser water temperature, the chiller can be completely turned off and the heat exchanger can be used to completely cool chilled water, or the chiller can be operated at reduced load by using the heat exchanger to precool the return water.

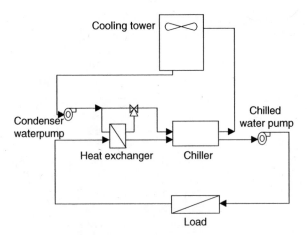

Figure 2.40   Arrangement of a waterside economizer.

2.4   Summary

In commercial buildings, chillers used for central air-conditioning systems are the biggest energy consumers. Therefore, significant energy savings can be obtained by improving the energy efficiency of chiller systems. This chapter provided an introduction to principles of refrigeration, refrigeration cycles, and types of chillers and how they operate. Thereafter, a considerable number of energy management strategies, which cover both design and operational aspects of chiller systems, were presented and explained using the theory of refrigeration systems. Various examples were also used to illustrate how the feasibility of implementing the different improvement measures can be evaluated by estimating the savings for each measure.

Review Questions

2.1. State four possible strategies to improve the efficiency of water-cooled chillers. For each strategy, briefly describe how it will result in an improvement in the chiller efficiency.

2.2. Briefly describe how variable speed drives (VSDs) can be used to reduce the energy consumption of chilled water pumps.

2.3. A 500-RT chiller operates 10 hours a day and has an operating efficiency of 0.6 kW/RT when the chilled water supply temperature is 7°C. If the chilled water supply temperature is increased to 8°C, estimate the kWh savings per day that can be achieved assuming the chiller efficiency improves 3 percent for every degree Celsius increase in chilled water supply temperature.

2.4. The rated efficiency and cost of three 500-RT capacity chillers are as follows:

Chiller	First cost	Efficiency (kW/RT)
1	$300,000	0.5
2	$275,000	0.55
3	$250,000	0.65

Compute the life-cycle cost for a 10-year period for each chiller based on the following:

Operating hours = 10 h/day and 250 days a year

Electricity tariff = $0.10/kWh

Electricity cost escalation = 2% a year

2.5. An office building uses a 150-RT chiller at night to provide after-office hours cooling to satisfy a load of 20 RT from 6 p.m. to 2 a.m. (8 hours). The resulting chiller efficiency is 1.75 kW/RT.

Calculate the daily kWh savings if the after-office hours cooling can be provided using a chilled water storage system where the chilled water can be produced using the day chillers operating at 0.65 kW/RT.

2.6. An industrial plant, which operates 24 hours a day, is considering replacing one of its 500-RT capacity chillers. Based on measurements carried out, the average daily electrical energy consumed by the chiller is 6000 kWh.

It is proposed to replace the existing chiller with a new chiller having the performance characteristic given in Table A.

Table A Performance Data for new Chiller

Chiller loading (RT)	Chiller efficiency (kW/RT)
500	0.55
400	0.56
300	0.57
200	0.6
100	0.75

The expected cooling load profile for the new chiller is given in Table B.

Table B Expected Cooling Load Profile

Time	Cooling load (RT)
12 a.m. to 6 a.m.	300
6 a.m. to 10 a.m.	400
10 a.m. to 2 p.m.	450
2 p.m. to 8 p.m.	400
8 p.m. to 12 a.m.	350

(i) Calculate the average daily energy savings (in kWh) that will result if the existing 500-RT chiller is replaced with the new chiller (use linear interpolation to obtain chiller efficiency data from Table A).

(ii) Payback period for replacing the chiller, assuming the chiller operates 300 days a year and the cost of the new chiller is $300,000 (you may ignore power demand savings).

3

Boilers and Heating Systems

3.1 Introduction

Boilers are pressure vessels used in buildings and industrial facilities for heating water or producing steam. They are primarily used for providing space heating for buildings in temperate climates as well as for producing hot water and steam required by users such as laundries and kitchens. For space heating, boilers function like chillers in central air-conditioning systems and provide steam or hot water to different parts of the building for heating. A typical arrangement of a boiler plant used for heating is shown in Fig. 3.1.

Boilers are either hot water boilers or steam boilers and are able to burn fossil fuels like oil, gas, and coal (some use electric current). Water boilers are normally low-pressure and are used primarily for space heating and producing hot water. Steam boilers are used for space heating as well as in other applications that require steam.

In facilities that use boilers, a large percentage of the energy (fuel) consumption is accounted for by the boiler plant. As such, significant energy savings can be achieved by optimizing boiler systems. This chapter describes some fundamental features of boilers and steam systems, followed by possible energy saving strategies for such systems.

3.1.1 Boiler construction

A boiler generally consists of a combustion chamber, which can burn fuel, in the form of solid, liquid, or gas, to produce hot combustion gases, and a tubular heat exchanger, to transfer heat from the combustion gases to the water.

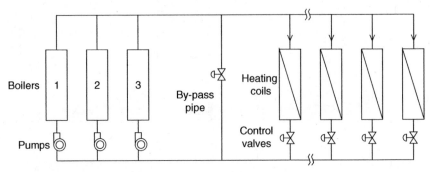

Figure 3.1 Typical arrangement of boiler plant used for space heating.

As shown in Fig. 3.2, the main inflows to a typical boiler are fuel, air, and feedwater while the outflows are steam or hot water, exhaust flue gases, and blowdown.

Boilers are normally classified as fire tube or water tube boilers, depending on the flow arrangement of water and hot gases inside the boiler. In fire tube boilers, the hot gases pass through boiler tubes that are immersed in the water being heated, while in water tube boilers, water is contained in the tubes that are surrounded by the hot combustion gases. Typical arrangements of fire tube boilers and water tube boilers are shown schematically in Figs. 3.3 and 3.4.

To increase the surface area available for heat transfer between the combustion gases and water, the tubes in boilers are arranged to have a number of passes so that the hot flue gases and water can pass through a number of sets of tubes before being exhausted. Cutaway pictures of the two types of boilers are shown in Figs. 3.5 and 3.6.

Boiler capacity depends on the steam generation rate and steam pressure. Steam generation is usually rated in kg/h, lb/h, or tons/h, while steam pressure is rated in psi (pounds per square inch) or bar.

Sometimes, boiler capacity is rated in boiler horsepower (BHp), where 1 BHp is equal to 9.803 kW or 3.3457×10^4 Btu/h (3450 lb/h = 100 Hp).

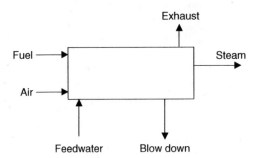

Figure 3.2 Main inflows and outflows for a typical boiler.

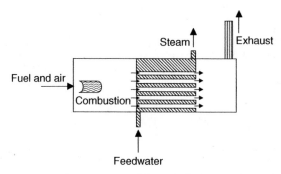

Figure 3.3   Arrangement of fire tube boilers.

3.1.2   Boiler efficiency

A typical heat balance for a boiler is shown in Fig. 3.7. As shown in the figure, only part of the heat content of the fuel is converted into useful heat, while the rest is lost through exhaust gases, blowdown, and radiation losses. The efficiency of boilers is usually rated based on combustion efficiency, thermal efficiency, and overall efficiency.

Combustion efficiency.   The typical combustion process in boilers involve burning of fuels that contain carbon (oil, gas, and coal) with oxygen to generate heat. Oxygen required for combustion is normally taken from air supplied to the burner of the boiler. The amount of air needed for combustion depends on the type of fuel used. To ensure complete combustion of fuel, more air than required (excess air) for combustion is provided to ensure that the fuel is completely burnt. Since excess air leads to lower boiler efficiency (due to removal of heat by the excess air as it passes through the boiler), the objective is to ensure that the optimum amount of excess air is provided.

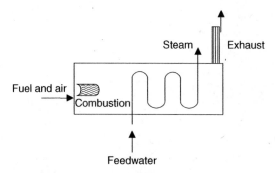

Figure 3.4   Arrangement of water tube boilers.

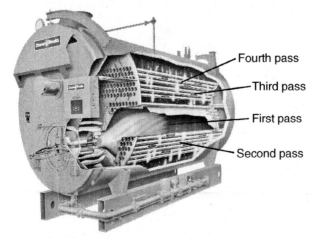

Figure 3.5 Cutaway of fire tube boiler. (*Courtesy of Cleaver Brooks.*)

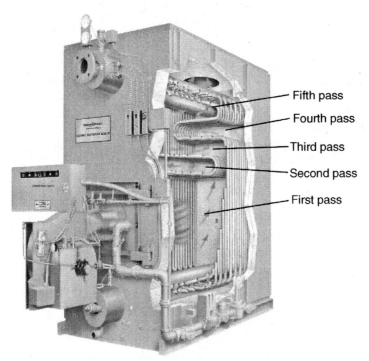

Figure 3.6 Cutaway of a commercial water tube boiler. (*Courtesy of Cleaver Brooks.*)

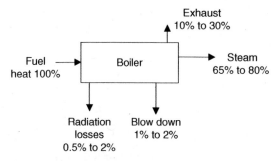

Figure 3.7   Typical heat balance for a boiler.

One of the most common measures of boiler efficiency is combustion efficiency, which indicates the ability of the combustion process to burn the fuel completely. It is normally measured by sampling the exhaust flue gas to find the composition and temperature using a combustion analyzer. Most good combustion analyzers are able to give a direct reading of the combustion-efficiency based on the fuel used. If this facility is not available on the instrument used, combustion efficiency charts available for different fuel types can be used to estimate the combustion efficiency.

Thermal efficiency.   Thermal efficiency is a measure of the efficiency of the heat exchange in the boiler. It provides an indication of how well the heat exchanger can transfer heat from the combustion process to water or steam in the boiler. It does not take into consideration the conduction and convection losses from the boiler.

Overall efficiency.   Another measure of boiler efficiency is the overall boiler efficiency, which is a measure of how well the boiler can convert the heat input from the combustion process to the steam or hot water. It is also called fuel-to-steam efficiency.

$$\text{Overall boiler efficiency} = \frac{\text{Heat output}}{\text{Heat input}}$$

The heat input depends on the amount of fuel burnt and its calorific value (heating value). The calorific value, normally expressed in kJ/kg, multiplied by the amount of fuel burnt in kg/s gives the heat input in kJ/s (kW).

The heat output is the difference in the heat content of the feedwater and steam (or hot water) produced multiplied by the flow rate of water or steam. The heat content of water and steam is expressed in kJ/kg and the flow rate of water or steam is expressed in kg/s, which yields the heat output in kW.

The overall efficiency of a boiler is lower than the combustion efficiency as it takes into account radiative and convective losses from the boiler and other losses, such as cycle losses, due to passing of air through the boiler during the "off" cycle.

The efficiency of a boiler can also be estimated by subtracting stack losses, radiative losses, and convective losses from the combustion efficiency. While combustion efficiency can be measured directly by using a combustion analyzer, the stack, radiative, and convective losses can be estimated using boiler manufacturers' data.

Table 3.1 shows the expected radiation and convection losses for a boiler while Tables A.1, A.2, and A.3 in Appendix A show the approximate stack losses (based on CO_2 concentration in flue gas and the difference in temperature between the flue gas and boiler room). If the combustion analyzer does not provide CO_2 concentration, Fig. A.1 in Appendix A can be used to convert O_2 values to CO_2 before using Tables A.1, A.2, and A.3.

Example 3.1 Consider a 150 BHp (1471 kW) boiler operating on natural gas at a 75 percent firing rate. The flue gas is sampled using a flue analyzer, which shows that the flue gas temperature is 200°C and the CO_2 level is 5 percent.

If the boiler room temperature is 35°C, the temperature difference between the flue gas and room temperature is 165°C (300°F).

Using the stack loss Table A.1 in Appendix A, the stack loss can be estimated to be 22.2 percent

If the boiler operating pressure is 6.9 bar (100 psi), the radiative and convective losses can be estimated to be 0.7 percent (Table 3.1).

Therefore, the overall fuel-to-steam efficiency = 100 − (22.2 + 0.7) = 77.1 percent.

3.1.3 Auxiliary equipment

Boilers need various auxiliary equipment such as feedwater pumps, draft fans, feedwater tanks, condensate recovery tanks, deaerators, and

TABLE 3.1 Approximate Radiation and Convection Losses for a 4-Pass Boiler Well Insulated for High Efficiency

Firing rate (% of load)	100–350 BHp (981–3433 kW)		400–800 BHp (3924–7848 kW)	
	Operating pressure 10 psig (69 kPa)	Operating pressure 125 psig (863 kPa)	Operating pressure 10 psig (69 kPa)	Operating pressure 125 psig (863 kPa)
25%	1.6%	1.9%	1.0%	1.2%
50%	0.7 %	1.0%	0.5%	0.6%
75%	0.5%	0.7%	0.3%	0.4%
100%	0.4%	0.5%	0.2%	0.3%

(*Courtesy of Cleaver-Brooks.*)

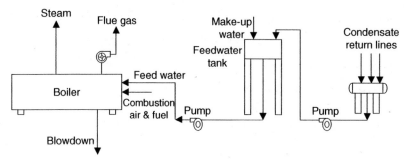

Figure 3.8 Typical boiler plant.

water softening plants for their operation. Figure 3.8 shows a typical arrangement of a boiler system with its main auxiliary equipment. The fan shown is for an induced draft arrangement although the actual operation can also be forced draft or natural draft. Some auxiliary equipment such as deaerators and water treatment systems are not shown in order to simplify the arrangement.

3.2   Energy Saving Measures for Boiler Systems

3.2.1   Improving combustion efficiency

The major loss in any boiler is due to the hot gases discharged into the chimney. If there is a lot of excess air, the increased quantity of exhaust gas will lead to extra flue gas losses. Similarly, insufficient air for combustion results in wastage of fuel due to incomplete combustion and reduces the heat transfer efficiency due to soot build up on heat transfer surfaces.

The amount of excess air required depends on the type of fuel and, in general, a minimum of about 10 to 15 percent excess air is required for complete combustion. This translates to about 2 to 3 percent excess oxygen.

Boiler combustion efficiency, which indicates the ability of the combustion process to burn the fuel completely (with minimum excess air), can be measured by sampling the exhaust flue gas to find its composition and temperature using a combustion analyzer. Most good combustion analyzers are able to give a direct reading of the combustion efficiency based on the fuel used. If this facility is not available on the instrument used, a combustion efficiency versus oxygen (O_2) concentration chart (Fig. 3.9) can be used to estimate the combustion efficiency.

The drop in combustion efficiency due to excess air is dependent on the type of boiler and the amount of excess air. Based on the chart

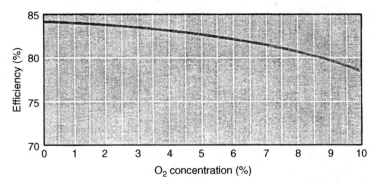

Figure 3.9   Combustion efficiency versus O_2 concentration. *(Courtesy of Cleaver Brooks.)*

(Fig. 3.9), if the excess air is increased from 15 to 30 percent, the O_2 concentration will increase from 3 to 5 percent (as normal air contains 21 percent oxygen) and the resulting drop in efficiency will be about 1 percent.

Example 3.2   If the flue gas from a boiler contains 10 percent oxygen, the amount of excess air can be estimated as follows:

It is necessary to imagine that there are two streams of air going into the boiler, one for complete combustion and the other to provide excess air. Since the oxygen in the combustion air will all be used up, only the oxygen from the excess air will be left in the flue gas.

Therefore,

Percentage of oxygen in air × excess air flow = total flue gas flow
× percentage of oxygen in flue gas

21 percent × excess air flow = total flue gas flow × 10 percent

Total flue gas flow/excess air flow = 2.1

This shows that the total flue gas flow is double the excess air flow i.e. excess air flow = combustion air flow (100 percent excess air).

In Fig. 3.9, if the excess air is reduced from 100 to 15 percent (O_2 concentration from 10 to 3 percent), the combustion efficiency is expected to increase by about 5 to 6 percent.

For boilers operating at high excess-air levels, the combustion burner operation needs to be tuned to adjust the air-to-fuel ratio. This can normally be achieved by adjusting the mechanical linkages that control fuel and air flow to the burner to provide the correct ratio between the two at different operating loads for the boiler. Ideally, an oxygen (O_2) trim system should be installed, which can continuously monitor the oxygen level in the flue gas and automatically adjust the air-to-fuel ratio to maximize combustion efficiency.

The amount of excess air also increases sometimes due to excessive draft created by the stack. If the stack is high, the natural draft created by the buoyancy of the combustion gases can be significant. This effect can be overcome by having a draft-control system, which consists of an opening with a damper, installed on the exhaust duct between the boilers and the stack, to automatically control the draft by opening or closing the damper.

The amount of excess air required for combustion also depends on the type of burner. Some old burners require much more excess air for complete combustion than others. Such burners can also be replaced with low excess-air burners to improve combustion efficiency.

3.2.2 Optimizing steam pressure

Boilers have a maximum operating pressure rating, based on their construction, as well as a minimum value to prevent carryover of water. The actual operating pressure is normally set based on the requirements of the end users, while ensuring it is within the specified maximum and minimum values.

Since boiler efficiency depends on operating pressure, if the operating pressure is set much higher than required, energy savings can be achieved by reducing it to match the actual requirements. Typically, reducing boiler pressure can help improve boiler efficiency by 1 to 2 percent.

In addition to improving boiler efficiency, reducing steam pressure helps to reduce steam leaks and wastage due to overheating in some applications. Reducing pressure also lowers the temperature of the distribution piping, which helps to cut down on losses. Another benefit of reducing pressure is the reduction of flash steam from vents of condensate recovery systems.

The heat carrying capacity (latent heat) of steam reduces with increase in pressure. Since many applications of steam involve condensing of steam in heat exchangers, it is best to keep the steam pressure at the lowest acceptable value to extract the maximum latent heat from steam.

However, it should be noted that when steam pressure is reduced, the distribution pipe sizing needs to be sufficient to transport the higher volume of steam.

If it is not possible to reduce the pressure of the entire system, parts of the distribution system can be operated at lower pressures by installing pressure reducing valves at appropriate points in the distribution network.

In some systems, one steam user may require steam at a much higher pressure than the others. In such a system, if the steam usage of the high-pressure user is relatively low, it may be better to have a separate steam generator (located near the user) operating at the required higher pressure while the rest of the system can be operated at a lower pressure.

Further, the system pressure requirements may vary at different times. For example, during daytime a higher pressure may be required to operate certain laundry equipment while at night a lower pressure may be sufficient for providing only space heating. In such a situation, it may be possible to have different set points for boilers during daytime and nighttime operations.

3.2.3 Fuel switching

Many modern boilers are the dual-fuel type and can operate on different fuels like fuel oil or gas. Since most of the cost of operating a boiler plant is accounted for by fuel cost, switching between fuels based on cost can help reduce energy cost.

The cost of fuels, such as oil and gas, can vary depending on seasonal factors and due to other reasons. Therefore, a particular fuel may not be the most economical to use at all times and switching between fuels can help minimize fuel cost.

The switch over from one fuel to another can be done manually by plant operators or automatically in some boilers. The decision on when to switch fuel should be used based on cost can be made by comparing the cost of fuel per unit heat content. The heat content of fuels vary depending on factors such as their composition, but the actual heat content of a particular fuel used can normally be obtained from the fuel supplier. The approximate heat content values of some typical fuels are listed in Table 3.2.

Example 3.3 A 400-BHp boiler requires 0.12 L/s of No. 2 oil when operating at 80 percent efficiency. The same boiler requires 0.13 m^3/s of natural gas when operating at the same efficiency.

If the cost of No. 2 oil is $0.30/L and the cost of natural gas is $0.18/m^3, the hourly operating cost for using No. 2 oil is 0.12 L/s × 3600 s/h × $0.30 /L = $129.60, and the hourly operating cost for using gas = 0.13 m^3/s × 3600 s/h × $0.18/ m^3 = $84.24. Based on this, it is cheaper to run the boiler on natural gas than on No. 2 oil.

TABLE 3.2   Approximate Heat Content Values for Fuels

Fuel	Approximate heat content (MJ/kg)
Natural gas	55
No. 2 oil (light oil)	46
No. 4 oil (heavy oil)	45
No. 6 oil (heavy oil)	44

Ideally, other factors such as boiler efficiency and maintenance costs need to be factored into the cost comparison as boiler efficiency and maintenance cost may vary depending on fuel use.

3.2.4 Optimizing operation of auxiliary equipment

Auxiliary equipment such as feedwater pumps, boiler draft fans, hot water circulating pumps, and condensate pumps also consume an appreciable amount of energy. Therefore, significant energy savings can be achieved by ensuring that they are operated only when required and at the capacity required to maintain system requirements.

In some installations that have additional equipment to provide extra reliability (standby equipment) or to match certain boiler load conditions, plant operators may run more equipment than required to meet the operating load. In such situations, some auxiliary equipment can be switched off either manually or by using automatic controls.

Generally, each boiler has its own feedwater pump, which is automatically switched on and off to maintain the level of water in the boiler. Their operation is interlocked with the boiler so that the feedwater pump is switched off when the boiler is not in operation.

In larger systems, multiple boilers can be served by a common set of feedwater pumps, as shown in Fig. 3.10. In such an arrangement, individual boilers take the required water flow by opening and closing the feedwater valves to maintain the water level in the boilers. The excess water is returned to the feedwater tank, which results in wastage of pumping energy.

This system can be improved to reduce the energy consumption of the pumps by varying the capacity (speed) of the feedwater pumps, which helps maintain a set pressure in the feedwater header pipe, as shown

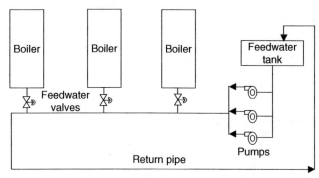

Figure 3.10 Feedwater pump arrangement for a multiple boiler operation.

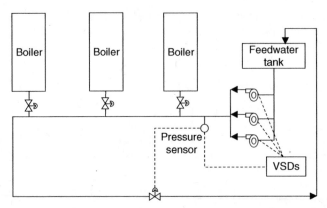

Figure 3.11 Suggested feedwater pump arrangement for a multiple boiler operation.

in Fig. 3.11. A pressure-activated valve (normally closed) can be installed on the return pipe as a safety measure, so that it opens if the pressure exceeds a set value (which may occur due to failure of the pump control system).

Boiler fans used to create the draft necessary for combustion and carry the flue gases through the boiler normally operate at constant speed and dampers are used to control the air flow to match boiler load conditions. In such systems, when the boiler operates at part load, a damper throttles the air flow by inducing a resistance across the path of the air flow. As a result, the energy consumption of the fan does not reduce proportionately to the air flow. However, if a variable speed fan is used for this application (Fig. 3.12), due to the cube law [fan power $\propto$ (air flow rate)^3], the reduction in fan energy consumption would be proportional to the third power of the load. Therefore, theoretically, if the load on the boiler reduces by 20 percent, the energy consumption of the fan will be reduced by about 50 percent (0.8^3).

The application of this energy saving measure depends on the load profile of the boiler. If the load is highly variable and results in the boiler operating at low loads for long periods of time, this is a good

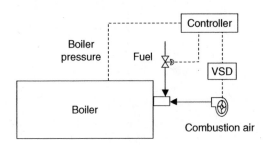

Figure 3.12 Application of VSD for boiler fan.

TABLE 3.3   Boiler Operating Data for Example 3.4

Boiler loading	Operating hours a day	Fan motor power (kW)
100%	2	22
80%	4	21
60%	10	19
40%	8	16

opportunity to incorporate a variable speed drive for the forced draft or induced draft fan of the boiler. Generally, such a retrofit is most economical in large boilers with modulating burners.

Installation of variable speed drives for boiler fans may require consultation with the boiler manufacturer to ensure that the necessary control modifications (to keep the damper fully open while controlling the fan speed based on load) can achieve the proper air-to-fuel ratio at different load conditions.

The savings achievable by using a VSD for boiler fan speed control can be estimated by recording the boiler operating load profile and the fan power consumption, as illustrated in Example 3.4.

Example 3.4   The operating loading and associated forced-draft fan power consumption of a boiler is given in Table 3.3.

If the boiler users a damper system to control the air flow rate, the energy savings that can be achieved by installing a VSD can be estimated as shown in Table 3.4.

Based on Table 3.4, the total savings is 296.8 kWh a day. This value can be multiplied by the number of operating days a year and the electricity tariff to calculate the annual cost savings.

3.2.5 Standby losses

Standby losses take place when a boiler is not firing and the hot surfaces inside the boiler lose heat to colder air circulating inside it. Such air circulation can take place due to natural convection and purging.

TABLE 3.4   Estimate of Savings for Example 3.4

Boiler loading A	Operating hours a day B	Fan motor power with damper (kW) C	Fan motor power with VSD (kW) $D = A^3 \times 22$	Power saving (kW) $E = C - D$	Energy savings (kWh) $F = B \times E$
100%	2	22	22	0	0
80%	4	21	11	10	40
60%	10	19	5	14	140
40%	8	16	1.4	14.6	116.8
				Total	296.8

Note: Column D is (boiler loading)^3 × fan power at full load which is 22 kW.

Losses due to natural convection occur when the air in the boiler gets heated (by the hot surfaces), making it lighter and causing it to moves up the stack circulating cold air through the boiler. This can be avoided if dampers are installed to prevent the circulation of air when the boiler is not being fired.

Purging losses take place when the boiler combustion space is purged by the fan before firing the burners to ensure that there is only air (to prevent possible explosions). Some burner systems also follow a purging cycle when firing stops. Losses due to purging can be reduced by minimizing the on-off cycle of the burner system. This can be achieved by using burners that have a high turndown ratio (ratio of maximum heat output to the minimum heat output of a burner) to enable the burner to function even at low loads without switching off the flame.

3.2.6   Minimizing conduction and radiation losses

Boilers, auxiliary equipment, and distribution piping of steam systems are much hotter than the surrounding areas. Therefore, they lose heat by radiation and conduction. The amount of heat lost depends on the surface temperature of the hot surface, which in turn depends on the insulation (thickness, thermal conductivity, and condition). To minimize heat loss, all hot surfaces should be insulated with material having sufficient resistance to heat transfer. Further, the insulation should be of adequate thickness and it should be in good condition.

Heat loss also depends on the area of the hot surface. Since boilers have large surface areas, heat loss from the boiler by radiation can be significant when operating at low loads. For a typical boiler operating at full load, heat loss due to radiation and convection is about 2 percent (Table 3.1 shows the losses for a boiler insulated for high efficiency). Since the radiative and convective losses remain the same irrespective of boiler loading, the 2 percent loss at full load can increase to 8 percent when the boiler is operating at 25 percent load (as illustrated in Table 3.5) for a typical boiler.

TABLE 3.5   Illustration of Radiation Losses for a Typical 200-BHp (1962 kW) Boiler

Boiler loading	Fuel input (kcal/h)	Radiation losses (kcal/h)	% Radiation loss
25%	500,000	40,000	8%
50%	1,000,000	40,000	4%
75%	1,500,000	40,000	3%
100%	2,000,000	40,000	2%

3.2.7   Preheating combustion air

In temperate climates, when boilers are used during winter to provide space heating, the fresh air drawn in for combustion can be quite cold. Since boilers and stacks release a considerable amount of heat into the boiler room, some of the warm air from the boiler room can be used for combustion. As warm air rises, resulting in stratification, the boiler air intake can be arranged to draw air from the higher levels of the boiler room. To make this effective, the ventilation openings at the higher levels of the boiler room need to be closed while ensuring that minimum ventilation requirements are maintained.

Typically, a 20°C increase in the temperature of combustion air can lead to a boiler efficiency improvement of about 1 percent.

3.2.8   Optimum start controls

Boiler systems are operated by manual controls or timers to start and stop them at fixed times. Such operations are scheduled based on routine requirements such as preparing to meet the building's heating requirements under the worst conditions, like the coldest period in the year.

Since the time taken to warm a building depends on factors like the air temperature of the space to be heated and the outdoor temperature, optimum-start controls, which use algorithms to predict the latest possible time to meet the system requirements based on the space temperature and outdoor temperature, can be used instead of manual or timer controls to minimize boiler operations.

Such a system would be able to start the boiler plant earlier on cold days or on Mondays after a weekend shutdown, or later on warmer days, while ensuring that the required space temperature is achieved when building occupancy begins.

Since boilers and their auxiliary equipment consume large amounts of energy to operate, minimizing their operating hours by an optimum-start strategy can help to significantly reduce energy consumption.

3.2.9   Heat recovery from flue gas

A significant amount of heat energy is lost through flue gases as all the heat produced by the burning fuel cannot be transferred to the water or steam in the boiler. As the temperature of the flue gas leaving a boiler typically ranges from 150 to 250°C, about 10 to 20 percent of the heat energy is lost through it.

Therefore, recovering part of the heat from flue gas can help to improve the efficiency of the boiler. Heat can be recovered from the flue gas by passing it through a heat exchanger (commonly called an economizer) installed after the boiler, as shown in Fig. 3.13. The recovered

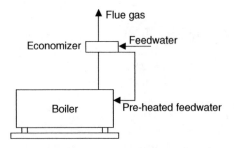

Figure 3.13 Arrangement of a typical economizer.

heat can be used to preheat boiler feedwater, combustion air, or for other applications. The amount of heat recovered depends on the flue gas temperature and the temperature of the fluid to be heated.

One of the major problems associated with flue gas heat recovery is corrosion due to acid condensation. Acid condensation takes place when the flue gas is cooled below its acid dew point. The sulfur in the fuel combines with water to form sulfuric acid which is corrosive. Therefore, the temperature of the flue gas needs to be maintained well above the acid dew point to prevent corrosion unless a heat recovery system specially designed to withstand acid corrosion is used.

The acid dew point depends on the sulfur content of the fuel. Some typical values are given in Table 3.6.

The feasibility of installing a heat recovery system for flue gas depends on factors such as by how much the stack temperature can be reduced, the inlet temperature of the fluid to be heated, and the operating hours of the boiler. Generally, the possible reduction in flue gas temperature should be at least 25 to 30°C to make it economically viable to install a heat recovery system.

Since economizers induce extra pressure losses on the flue gas and the liquid being heated, care should be taken to ensure that the combustion fan and the pump for the liquid being heated have adequate capacity to overcome these losses.

Example 3.5 A 4000 kg/h (4 ton/h) boiler using approximately 167 L/h of low sulfur oil operates with a flue gas temperature of 190°C. Find the energy savings possible if an economizer is installed to preheat feedwater at 90°C.

TABLE 3.6   Acid Dew point for Common Fuel Types

Fuel	Acid dewpoint temperature (°C)	Allowable exit stack temperature (°C)
Natural gas	66	120
Light oil	82	135
Low sulfur oil	93	150
High sulfur oil	110	160

For low sulfur oil, since the minimum allowable stack temperature is 150°C, the reduction in temperature possible for the flue gas is 40°C (190–150°C).

The amount of fuel used = 167l/h ≈ 150 kg/h (assuming the density to be 900 kg/m^3).

For perfect combustion, with just enough air for complete combustion, the air-to-fuel ratio is about 1:15 (stoichiometric mixture).

Therefore, the amount of combustion air is approximately 15 times the weight of the fuel used. i.e. 15 × 150 = 2250 kg/h (would be slightly higher based on the amount of excess air).

Total mass of flue gas = (2250 + 150) = 2400 kg/h = 0.67 kg/s.

Taking the specific heat capacity of flue gas to be 1.1 kJ/kg.K, the amount of heat recovered can be estimated as follows:

Heat recovered = mass flow rate × specific heat capacity

× temperature drop for flue gas

= 0.67 × 1.1 × 40

= 29 kW (kJ/s)

= 0.029 × 3,600 = 104.4 MJ/hr.

If the heat content of the fuel is 40 MJ/kg,
the reduction in fuel usage = 104.4/40 = 2.6 L/h.

The increase in the feedwater temperature can be estimated as follows:

Heat recovered = 29 kW = mass flow rate × specific heat capacity

× temperature rise for water

= (4000/3600) kg/s × 4.18 kJ/kg.K × (T – 90°)

From the above equation, the temperature of feedwater leaving the economizer (T) can be computed to be 96°C (an increase of 6°C).

3.2.10 Automatic blowdown control and heat recovery

Boiler blowdown is part of the water treatment process and involves removal of sludge and solids from the boiler. Makeup water used for boilers contain various impurities. As water is converted to steam, the concentration of the impurities that remain in the boiler increases. If this concentration is allowed to increase, it will lead accelerated corrosion, scaling, and fouling of the heat transfer surfaces of the boiler. Therefore, it is necessary to remove part of the concentrated water from the boiler and replace it with fresh water.

Boiler blowdown can be intermittent, where a fixed quantity of water is drained periodically, or continuous, where a small amount of the water is removed continuously to maintain the quality of water within acceptable limits.

Blowdown involves discharge of water at steam temperature, which has to be replaced by an equivalent amount of cold water. Energy losses resulting from blowdown can be minimized by installing automatic blowdown systems and recovering heat from blowdown.

Automatic blowdown control systems monitor the pH and conductivity of the boiler water and allow blowdown only when required to maintain an acceptable level of water quality. Heat recovery from blowdown involves utilizing a heat exchanger to preheat cold makeup water using the blowdown. Such systems are feasible for boilers that operate most of the year using at least 5 percent of makeup water. For high-pressure systems having steam pressures over 300 psi, flash steam can also be recovered from the blowdown and can be used as low-pressure steam or condensed back as part of the boiler feedwater. A typical arrangement of a boiler blowdown system with heat recovery is shown in Fig. 3.14.

3.2.11   Boiler operating configuration

The efficiency of boilers vary at different load conditions depending on how well the burner system can match load variations. Burner systems on boilers normally use single-stage, two-stage, or modulating type burners to vary the boiler output and match load requirements.

Single-stage burners have only one output setting and vary burner output by switching the burner on and off. This can lead to high standby losses, as explained earlier. Two-stage burners have low-fire (about 60 percent of maximum) and high-fire output (maximum) settings. In modulating burners, the heat output is modulated between a set of maximum and minimum settings to match load requirements. The minimum output can be as low as 25 percent of the maximum output for modulating burners.

Therefore, a sequencing program needs to be used for systems with multiple boilers to optimize their operation. The system should be able to minimize the heat input for a given steam load. For example, for a two-stage

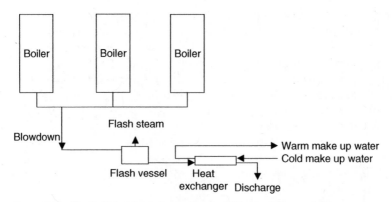

Figure 3.14   Typical blowdown flash steam and heat recovery system.

TABLE 3.7   Boiler Operation

Boiler no.	Boiler size $(10^3$ kg/h)	Boiler operating load $(10^3$ kg/h)	Heat input $(10^6$ kcal/h)
1	100	70	47
2	100	70	51
3	50	30	24
		170	122

burner, the system should be able to decide whether to fire a boiler on its second stage or start another boiler on the first stage when the load increases. Similarly, if the load is 120 percent of the capacity of one boiler, whether to operate two boilers at equal load (60 percent load each) or operate one at a higher load. To optimize in this manner, data on the operating efficiency of the boilers at different load conditions need to be available.

Example 3.6   Consider the following situation (Table 3.7) where three boilers have to supply a load of 170,000 kg/h and each has a different operating efficiency (Table 3.8).

Based on operating capacity and efficiency data for the boilers, the required 170,000 kg/h of steam can be met by operating a minimum of two boilers (boilers 1 and 2).

By trial and error, it can be found that operating boilers 1 and 2 at equal load (85,000 kg/h each) results in the lowest fuel input (118×10^6 kcal/h). Alternatively, if boiler 1 is loaded to 100 percent and boiler 2 is operated at 70 percent load, the resulting fuel input will be higher (118.5×10^6 kcal/h).

3.2.12   Condensate recovery

In most steam systems, steam is used mainly for heating by extracting its latent heat. The resulting condensate is at steam temperature and

TABLE 3.8   Boiler Operating Efficiency Data

Boiler no.	Steam load $(10^3$ kg/h)	Combustion efficiency (%)	Output $(10^6$ kcal/h)	Fuel input $(10^6$ kcal/h)
1	100	85	66	67
	85	86	48	56
	65	84	37	44
	50	81.5	28	35
2	100	77.5	57	73
	85	78	48	62
	65	77	37	48
	50	74	28	38
3	50	78	28	38
	40	78.5	24	31
	30	76.5	18	24
	25	73.5	14	19

still contains a considerable amount of heat energy. If steam is used at 100 psi (690 kPa), then the condensate contains about 25 percent of the heat used to produce steam and will be lost if the condensate is not returned to the system. Therefore, returning condensate to the boiler feedwater tank will result in significant fuel energy savings.

Since condensate is distilled water, it is ideal for use as boiler feedwater. Therefore, condensate recovery helps to reduce water consumption (water cost), water treatment cost, and blowdown.

Usually, a low feedwater temperature or high makeup water flow indicates that less condensate is recovered. If the makeup water flow is metered, in applications that do not consume live steam (such as open sparge coils and direct steam injection systems), the difference between the amount of steam produced and makeup water flow will give an indication of the amount of condensate that is not recovered.

Similarly, temperature measurement of feedwater, condensate return, and makeup water streams can also help to estimate the amount of condensate recovered, as illustrated in Example 3.7.

Example 3.7   Consider the following situation where feedwater is provided to a boiler at 60°C from the feedwater tank. Temperature of condensate water returning to the tank is 88°C, while the temperature of makeup water is 27°C.

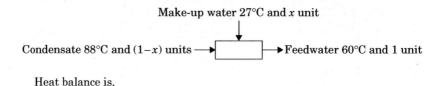

Make-up water 27°C and x unit

Condensate 88°C and $(1-x)$ units ⟶ [ ] ⟶ Feedwater 60°C and 1 unit

Heat balance is,

$$27x + (1 - x)\, 88 = 60$$

Therefore, $x = 0.36$ (36 percent makeup water or only 64 percent of condensate is recovered).

In some applications, condensate is not recovered due to possible contamination from leaking steam coils or heat exchangers, which can cause damage to boilers. This can be prevented by first returning the condensate from such applications to a collection tank fitted with a sensor to detect contamination before transferring to the main condensate tank. If this is not possible, for applications with large amounts of condensate, the heat can at least be recovered using a heat exchanger.

3.2.13   Steam traps

Steam traps are used in steam systems to remove condensate and noncondensable gases. They are mainly used in buildings for steam

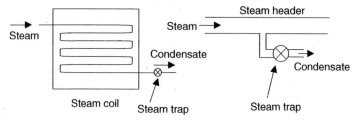

Figure 3.15 Application of steam traps.

heating coils and for condensate removal from steam headers, as shown in Fig. 3.15.

There are many types of steam traps (Fig. 3.16). Steam traps are generally classified as thermostatic, mechanical, or thermodynamic. Thermostatic steam traps are designed to work based on the difference in temperature of steam and condensate. They contain a bimetallic strip

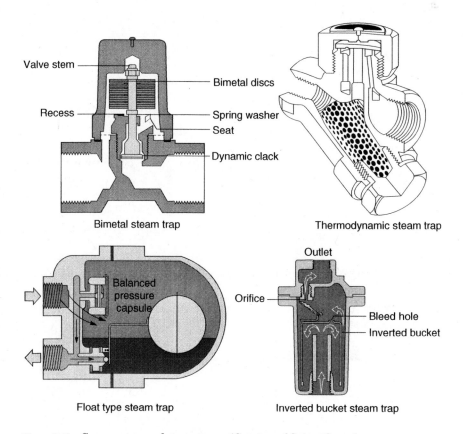

Figure 3.16 Common types of steam traps. (*Courtesy of Spirax Sarco.*)

or bellows to allow subcooled condensate to be removed while preventing live steam, which is at a higher temperature, from passing through. While bellows-type thermostatic traps can be used for steady light loads on low-pressure systems, bimetallic traps can be used for jacketed piping, steam tracers, and heat transfer equipment, which can accommodate backup condensate.

Bucket- and float-type traps are common types of mechanical steam traps. As the names imply, they have floating balls or buckets that operate on the buoyancy of condensate to mechanically open and close ports in the traps to discharge only condensate. They usually have built-in air venting features and are used for continuous and intermittent loads. They are commonly used on steam-heat exchanger coils.

Thermodynamic traps operate based on the difference in flow characteristics of steam and condensate. When air or condensate enters the trap, a disc lifts up to allow it to be discharged. When steam enters the trap, due to its increased velocity (higher velocity pressure), the static pressure below the disc is reduced, which lowers the disc, closing the trap. They are often used for condensate removal from main steam distribution pipes.

The operation of steam traps is important because if they fail to operate properly and allow live steam to pass through them from the steam side to the condensate side, it results in obvious loss of energy. In addition, if the traps are unable to remove air at start-up times or if they are unable to remove condensate at a sufficient rate, the resulting reduced capacity and longer periods to heat up would also result in energy wastage.

Over time, internal parts of steam traps wear out and result in failure to open and close properly. While an open trap would result in loss of live steam, a closed trap could result in loss of heat transfer area and water hammering. Water hammering can eventually result in damage to valves and other components in steam systems, which could result in steam leaks.

However, one of the main problems in maintaining steam traps is identifying defective steam traps. Often, the condensate released by traps is diverted to a condensate collecting tank, making it hard to spot leaking traps. Further, it is sometimes hard to distinguish between leaking steam and flash steam at the steam traps.

One way of identifying steam leaks from traps that are connected by piping to condensate tanks is to install sight glasses after the traps to facilitate visual indication of leaks. Ultrasound leak detectors can also be used to detect leaking traps. Further, traps should be periodically inspected and repaired or replaced to ensure that they are in good working condition. In addition, the correct type of steam trap should be selected for each application.

3.2.14 Steam leaks

Steam leakage occurs from pipes, flanges, valves, connections, traps, and process equipment and can be substantial for some steam distribution systems. The amount of steam leaking from various openings depends on the size of the opening and the system pressure. Figure 3.17 shows the approximate steam leak rates at different operating pressures for various leaking-hole sizes. The figure also gives the approximate fuel savings that can be achieved if a particular leak is eliminated based on boiler operation of 8400 and 2000 hours a year. The fuel savings for other annual operating hours can be estimated by interpolation.

Example 3.8 If the operating steam pressure is 600 kPa (6 bar) and the hole size is 7.5 mm, drawing a horizontal line on the chart through the point where the 600 KPa vertical line and the curve for the 7.5-mm hole intersect shows that the approximate steam leak rate is 110 kg/h.

If the boiler operates on heavy fuel oil, extending the horizontal line shows that the saving in fuel will be 70,000 L/year and 17,000 L/year, if the boiler operates 8400 and 2000 hours a year, respectively. Therefore, if the boiler operates 5000 hours a year, by linear interpolation, the approximate annual fuel savings is 42,000 L/year.

If the cost of fuel is $0.5/L, the annual cost savings that can be achieved by eliminating the leak is $21,000 (42,000 × 0.5).

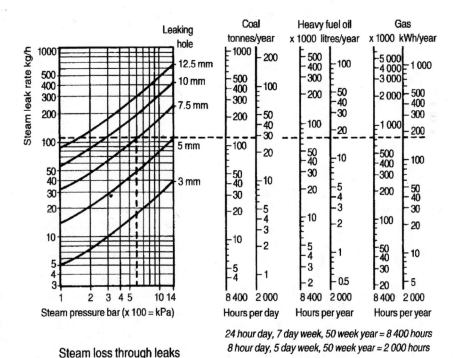

Steam loss through leaks

Figure 3.17 Steam losses through leaks. (*Courtesy of Spirax Sarco.*)

3.2.15   Feedwater tank

The feedwater tank is a very important part of any steam system. It provides a reservoir of returned condensate and fresh makeup water for the boilers. The feedwater tank gives a good indication of the system's health. Excessive feedwater temperature may indicate that some traps may be passing live steam, while a high makeup water flow may indicate that some condensate is not being returned to the tank.

Feedwater is normally hot due to returned condensate and recovery of heat from other sources. Therefore, the tank should be elevated to avoid hot water being flashed off as steam at the feedwater pump inlet, in order to prevent cavitation.

Since the feedwater tank is hot, steps should be taken to minimize heat losses from the tank. Other than insulating the tank, since a great amount of losses usually take place at the water surface, the top of the tank should be covered.

3.2.16   Fouling and scaling of boiler heat transfer surfaces

Fouling, scaling, and soot build up on heat transfer surfaces of boilers act as insulators and lead to reduced heat transfer. This results in lower heat transfer to the water in the boiler and higher flue gas temperature. If at the same load conditions and same excess air setting the flue gas temperature increases with time, this is a good indication of increased resistance to heat transfer in the boiler. Typically, 1- to 1.5-mm soot build up on the fire-side can increase fuel consumption by about 3 to 8 percent. Similarly, for the water-side, scale build up of 1 to 1.5 mm can result in extra fuel consumption of 4 to 9 percent.

When this occurs, the boiler heat transfer surface should be cleaned. On the fire-side, surfaces should be cleaned of soot, while on the water-side, scaling and fouling should be removed. For boilers using gas and light oil, it is generally sufficient to clean fire-side surfaces once a year. However, for boilers using heavy oil, cleaning may need to be done several times a year.

In addition, preventive steps should also be taken. For scaling, as it is caused by inadequate water treatment, steps should be taken to improve water softening and maintaining a lower total dissolved solids (TDS) level. For soot build up, which is normally due to defective burner or insufficient air for combustion, steps should be taken to repair or retune the combustion system.

3.2.17   Isolating off-line boilers

Sometimes, during light-load conditions, one boiler is used to meet the demand while other boilers are kept on standby. Such idling boilers can lose heat due to flow of air through the boiler into the stack. This can be avoided by installing dampers to automatically isolate the off-line boiler from the operational boilers and the stack.

3.2.18 Decentralized boiler systems

Usually, steam is generated by a central boiler and then distributed to the different areas of the building complex. In the case of large campuses, energy savings may be achieved by reducing distribution losses by having a decentralized system with smaller package boilers. This also enables the package boilers to be operated at different steam pressures, depending on the requirements of the individual areas, and will help lower the steam pressure.

3.2.19 Boiler replacement

Generally, the overall efficiency of boilers is similar. However, some boilers may operate at much lower efficiency than others due to reasons such as loading and characteristics of the burner. If a boiler is operated at 10 to 15 percent of its capacity, radiation losses can account for a significant percentage of the heat input from the fuel, resulting in low boiler efficiency. In such situations, it may be financially justifiable to replace the existing boiler with a new boiler that is sized to match the low operating load conditions.

If on the other hand, boiler efficiency is low due to the burner cycling on and off, due to low operating load leading to high standby losses caused by convection and purging, replacing the burner with a modulating burner with a high turndown ratio may help to improve boiler efficiency.

3.3 Ideal boiler system

A possible arrangement for a boiler system, incorporating some of the important energy saving measures discussed earlier in this chapter is shown in Fig. 3.18.

3.4 Heat Recovery Systems

Space heating and production of hot water can also be achieved using heat recovery systems to extract heat from waste heat sources. In buildings, the most common sources of waste heat are condensers of air-conditioning systems. In a typical air-conditioning system, the evaporator provides useful cooling while the condenser rejects the absorbed heat plus the heat added by the compressor to the environment. In such a system, rather than wasting this heat energy, the heat released by the condenser can be used for space heating or to produce hot water.

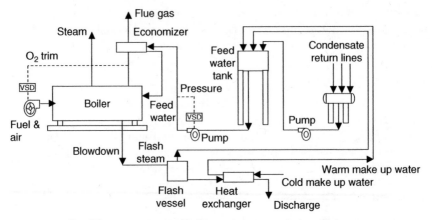

Figure 3.18   Possible arrangement of boiler system to maximize efficiency.

In typical heat recovery systems used with air-conditioning systems, the hot refrigerant is first passed through a heat exchanger before the condenser to extract heat (Fig. 3.19). Some of the common heat exchanger arrangements used are coils immersed in hot water tanks, refrigerant coils wrapped around hot water tanks, and plate heat exchangers. Since the refrigerant pressure is normally higher than the pressure of the water being heated, when using immersed coils and heat exchangers, a double-walled heat exchanger surface is needed to prevent contamination of the water in the event of a refrigerant leak.

In addition to recovering waste heat from the air-conditioning system, a heat exchanger also helps to reduce energy consumption by the condenser fan as most of the heat will be released in the heat recovery heat exchanger.

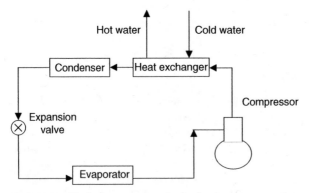

Figure 3.19   Typical arrangement of a heat recovery system coupled to an air-conditioning system.

3.5 Hydronic Heating Systems

As described earlier in the chapter, boilers commonly provide space heating for buildings by providing hot water from central plants, like chiller plants used for cooling. In these systems, which are sometimes called hydronic heating systems, hot water is pumped from the central plant to terminal units that have heating coils to provide space heating.

Energy savings achievable from such hydronic systems normally relate to optimizing hot water pumping systems, air distribution systems, and operating strategies. The optimizing measures for heating and cooling hydronic systems are similar and are therefore described together in Chapter 4 (pumping systems), Chapter 6 (air distribution systems), and Chapter 9 (building automation systems).

3.6 Summary

Boilers are used for providing space heating for buildings in temperate climates as well as for producing hot water and steam required by other users such as laundries and kitchens. Most boilers operate on fuel and are generally the biggest users of fuel in buildings. Therefore, significant savings in fuel consumption can be achieved by improving the efficiency of boilers and ancillary equipment used in boiler plants. This chapter provided an overview of boilers and how they operate. Thereafter, potential energy saving measures, which can improve the efficiency of boilers through operational strategies and system design, were described. Other opportunities for improving the efficiency of boiler systems, including optimization measures for ancillary equipment were also described in detail.

Review Questions

3.1. The flue gas from a boiler contains 8 percent oxygen. If boiler tuning is performed, the oxygen concentration in the flue gas can be reduced to 5 percent. Estimate the improvement in combustion efficiency that can be achieved by doing this.

3.2. A 200-BHp (1962 kW) boiler operates on No. 2 oil at a 75 percent firing rate. The flue gas is sampled using a flue analyzer, which shows that the flue gas temperature is 180°C and the CO_2 level is 5.5 percent. If the boiler room temperature is 38°C, estimate the overall boiler operating efficiency.

3.3. The operating load and associated forced-draft fan power consumption of a boiler is given below.

Boiler loading	Operating hours a day	Fan motor power (kW)
80%	4	22
60%	12	18
40%	8	16

If the boiler users a damper system to control the air flow rate, and the fan consumes 35 kW of power when the boiler operates at 100 percent loading, estimate the daily energy savings that can be achieved if a VSD is installed to control the fan capacity based on boiler load.

3.4. A boiler is operated to provide steam to calorifiers and to a laundry. Only part of the steam is returned as condensate to the system.

Feedwater is provided to the boiler at 80°C from the feedwater tank. The temperature of condensate water returning to the tank and that of the makeup water are 95 and 27°C, respectively. Estimate the amount of condensate that is recovered in the system.

3.5. A boiler operating on heavy fuel oil provides steam 24 hours a day to a distribution system that operates at 300 kPa pressure. If there are three holes of approximately 3 mm each in the distribution system that are leaking steam, estimate the annual fuel cost savings that can be achieved if the leaks are eliminated. Take the fuel cost to be $0.5 a liter.

4

Pumping Systems

4.1 Introduction

Various pumping systems are used in buildings. The most common systems are those used for pumping chilled water and condenser water in central air-conditioning systems and for pumping hot water in central heating systems. Pumping systems are also used in buildings for domestic hot water and cold water supply. This chapter deals mainly with pumping systems used in central air-conditioning and heating systems since they account for most of the energy consumed by pumping systems in buildings. However, some of the proposed energy management strategies may also be applicable to other pumping systems.

In chilled water and hot water (heating) systems, pumps are used to provide the primary force to distribute and circulate cold or hot water through the coils while overcoming pressure losses caused by the different components in the system. Similarly, in condenser cooling systems, water is circulated between the condensers and the cooling towers.

Centrifugal pumps are the most common type of pumps used in buildings. Centrifugal pumps have an impeller mounted on a shaft, which is driven by a motor and rotates in a volute or diffuser casing. In pumps with volute casings, water from the impeller is discharged perpendicular to the shaft, while in pumps with diffuser casings (in-line pumps), water is discharged parallel to the shaft.

Pumps are generally classified according to their installation arrangement and mechanical features. The most common pumps used in buildings are end-suction type pumps, which are horizontally mounted with single-suction impellers, horizontally- or vertically-split case pumps with double-suction impellers, and vertically mounted in-line pumps. (Figs. 4.1 and 4.2).

Figure 4.1   End-suction and in-line pumps. (*Courtesy of ITT Industries.*)

4.2 Distribution Systems

Chilled water and heating systems are closed systems where water is circulated in a closed loop. Condenser water systems are open systems where static pressure is present due to height difference when the water is sprayed in open cooling towers. Arrangements of closed and open systems are shown in Fig. 4.3.

The two main piping systems used for water circulation are direct return and reverse return systems. As shown in Figs. 4.4 and 4.5, the difference between direct return and reverse return systems is that in the latter, the water leaving the individual coils is combined together before returning to the main return header, whereas in the direct return system water return pipes are individually connected to the main return header.

As a result, in reverse return systems, the piping lengths between the circulating pumps and each coil is equal. Therefore, if the coils are selected to have the same water pressure drop, the system will be self-balancing and will eliminate the need for balancing valves. In direct

Figure 4.2   Horizontally- and vertically-split case pumps. (*Courtesy of ITT Industries.*)

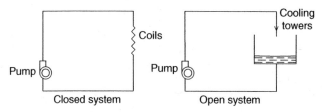

Figure 4.3   Arrangement of closed and open systems.

return systems, the piping length between coils and circulating pumps vary according to the location of the coils, with the nearest coil having the shortest pipe length. Direct return systems, therefore, need to have balancing valves on each branch of piping to prevent more water passing through the coils close to the circulating pump, leading to insufficient flow to the coils furthest away from the pump.

In addition to these water distribution systems, another variation, the primary-secondary pumping system, is also sometimes used. As shown in Fig. 4.6, a primary-secondary distribution system utilizes two sets of pumps. The first set of pumps, the primary pumps, is used to pump water through the chillers or boilers. The second set of pumps, the secondary pumps, is used to pump water through the coils located in the different parts of the building. The primary and secondary pumps are hydraulically isolated from each other by a bypass pipe called the decoupler pipe. The secondary distribution can be configured to be direct return or reverse return.

The flow in the decoupler piping can be in either direction, depending on the "production" of chilled or hot water in the primary circuit and the amount of "consumption" by the building (secondary flow). If the primary system produces more chilled or hot water than what the secondary system consumes, the flow of water in the decoupler pipe will be

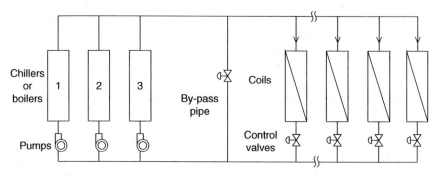

Figure 4.4   Direct-return distribution system.

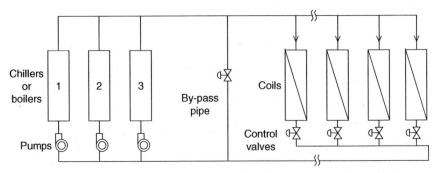

Figure 4.5 Reverse return distribution system.

from supply header to return header. On the other hand, if the secondary system requires more water than that produced by the primary system, the flow of water in the decoupler pipe will be from return to supply.

In such primary-secondary systems, hydraulic isolation allows the secondary pumps to vary the flow (usually using variable speed drives) with building load while maintaining a constant flow of water through the primary circuit. The advantages of such a system will be discussed later in the chapter.

4.3   System and Pump Curves

Pumps and pumping systems are normally rated based on the pressure head and flow rate. The two parameters are dependent on each other as the flow rate in a pumping system depends on the pressure head.

The pressure developed by a pump is necessary to overcome resistances in the system, such as those due to frictional losses in piping, pressure

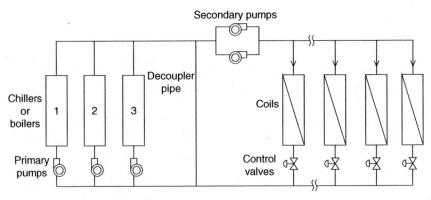

Figure 4.6 Primary-secondary pumping system.

losses across valves, and cooling coils and static head differences in open systems. The relationship between head losses in a system to the system flow rate is called the system resistance curve. Typical system curves for closed and open systems are shown in Fig. 4.7.

The difference between the two curves is that for open systems the static pressure difference or independent pressure due to height difference is added to the system curve. The system curve is parabolic in shape since the pressure losses in the system are proportional to the square of the flow ($\Delta P \propto \text{Flow}^2$).

Pressure drop due to friction of a fluid flowing in a pipe is given by the Darcy-Weisbach equation:

$$\Delta h = f \frac{L}{D} \frac{V^2}{2g} \tag{4.1}$$

where Δh = friction loss, m
$\quad f$ = friction factor, dimensionless
$\quad L$ = length of pipe, m
$\quad D$ = inside diameter of pipe, m
$\quad V$ = average velocity of fluid, m/s
$\quad g$ = acceleration due to gravity, 9.8 m/s^2

This has the relationship (pressure $\propto \text{Flow}^2$).

Further, pressure losses in fittings are also proportional to the square of the flow and can be expressed as:

$$\Delta h = K \frac{V^2}{2g} \tag{4.2}$$

where, K = loss coefficient, depending on type of fitting, size and flow velocity.

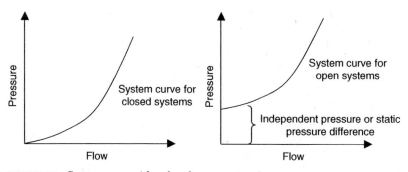

FIGURE 4.7   System curves (closed and open systems).

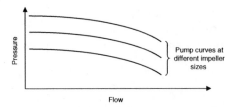

Figure 4.8 Pump curve (different diameters).

Each particular system will have its own system curve due to its unique pipe sizing, pipe length, and fitting. The system curve will, therefore, change if components in the system are changed; for example, with those having different flow resistances.

Similarly, the relationship between flow rate and pressure developed by a pump is called a pump curve. The pump curve shows all the different operating points of a pump at a particular operating speed as its discharge is throttled from zero to full flow. Since pumps can operate at different speeds and with different impeller sizes, usually pump curves are plotted on the same axis. Figure 4.8 shows the pump curves for a particular pump using different impeller sizes.

Such pump curves provided by pump manufacturers usually show not only the relationship between flow rate and pressure, but also the pump power and operating efficiency. The pump curves can be flat or steep, as shown in Fig. 4.9. In pumps with flat curves, large variations in flow can be achieved with relatively less change in pressure. Therefore, pumps with flat curves are preferred for closed systems with modulating 2-way control valves. For constant flow applications such as condenser water systems serving cooling towers, pumps with a steep characteristic can be used.

When a pump is selected for a particular application, a pump with a performance curve that can intersect the system curve at the desired operating point is selected, as shown in Fig. 4.10.

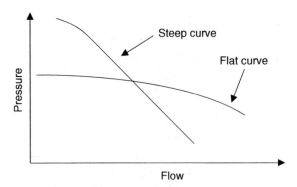

Figure 4.9 Flat and steep pump characteristics.

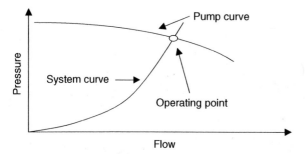

Figure 4.10 System and pump curve for a pumping system.

The power consumed by a pump is related to the product of the flow and the pressure difference (between discharge and suction), that is,

$$\text{Pump power} \propto \frac{\text{liquid flow rate} \times \text{pressure difference}}{\text{efficiency}} \quad (4.3)$$

In SI units:

Pump impeller power (kW)

$$= \frac{\text{flow rate (m}^3\text{/s)} \times \text{pressure difference (N/m}^2)}{1000 \times \text{efficiency}} \quad (4.4)$$

In Imperial units:

Pump brake horsepower

$$= \frac{\text{flow rate (USgpm)} \times \text{pressure difference (ft water)}}{3960 \times \text{efficiency}} \quad (4.5)$$

Therefore, pumping power consumption can be lowered by reducing the flow rate, the pressure difference, or both.

4.4 Affinity Laws

The performance of centrifugal pumps under different conditions are related by the pump affinity laws given in Table 4.1. The pump affinity laws relate pump speed and impeller diameter to flow, pressure developed across the pump, and brake horsepower of the pump.

Pump affinity laws are useful for estimating pump performance at different rotating speeds or impeller diameters, based on a pump with a known relationship. Example 4.1 illustrates the use of pump affinity laws.

TABLE 4.1   Pump Affinity Laws

	Change in speed (N)	Change in impeller diameter (D)
Flow (Q)	$Q_2 = Q_1\left(\dfrac{N_2}{N_1}\right)$	$Q_2 = Q_1\left(\dfrac{D_2}{D_1}\right)$
Pressure (Δp)	$\Delta p_2 = \Delta p_1\left(\dfrac{N_2}{N_1}\right)^2$	$\Delta p_2 = \Delta p_1\left(\dfrac{D_2}{D_1}\right)^2$
Power (P)	$P_2 = P_1\left(\dfrac{N_2}{N_1}\right)^3$	$P_2 = P_1\left(\dfrac{D_2}{D_1}\right)^3$

Example 4.1   A pump delivers 120 L/s at 1400 rpm and consumes 55 kW. If the pump speed is reduced to 1120 rpm, calculate the new flow rate and power consumption.

$Q_1 = 120$ L/s

$P_1 = 55$ kW

$N_1 = 1400$ rpm

$N_2 = 1120$ rpm

$$Q_2 = Q_1 \times (N_2/N_1) = 120 \times (1120/1400) = 96 \text{ L/s}$$

$$P_2 = P_1 \times (N_2/N_1)^3 = 55 \times (1120/1400)^3 = 28 \text{ kW}$$

Example 4.1 shows that when the speed is reduced by 20 percent (1400 rpm to 1120 rpm), theoretically, the power consumption reduces by about 50 percent (55 kW to 28 kW). However, it should be noted that the pressure developed by the pump also reduces to the square of the reduction in speed.

This is the well known "cube law" (power $\propto$ speed^3), which is very useful for energy savings in pumping systems, and its applications will be described later.

4.5 Energy Saving Measures for Pumping Systems

4.5.1 Pump sizing

Pumps are sized to take care of the design flow requirements while overcoming the various resistances in the system. Friction losses in piping and losses across valves and fittings are normally estimated using specification and research data. Due to the uncertainty of these estimated values and provision for possible changes during installation

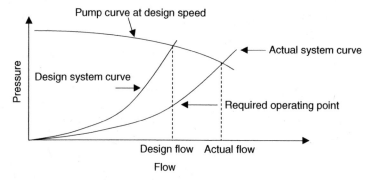

Figure 4.11 Over sized pump for application.

to suit site constraints, safety factors are added to the design. The safety factor can range from a few percent to as high as 100 percent. As a result, a pump can end up being over sized for the particular application, as illustrated in Fig. 4.11.

Figure 4.11 illustrates a case where the system curve used at the design stage has a high safety factor incorporated into it. The pump is selected to intersect this design system curve at the design flow to give the design operating point. However, since the design system curve has a high safety factor, the actual system curve the pump experiences may be quite different. This results in the pump operating point moving along the pump curve to where it intersects the actual system curve, leading to a higher than required pump flow rate.

Since the pumping power is related to the cube of the flow rate, overpumping by 20 percent results in 50 percent increase in pumping power consumption. For chilled water, overpumping can also lead to increase in the chilled water supply temperature since chillers are unable to provide chilled water at the design value when the flow rate is exceeded. This can result in less moisture removal at the terminal units, since the moisture removal ability of cooling coils is dependent on the coil temperature.

Usually, when pumps are oversized, they are either operated to give higher than required flow rates or a pressure loss is artificially created in the system by adding a throttling valve. Usually, globe valves or balancing valves are used in the system to add sufficient resistance to the system to move the actual system curve so that it intersects the pump curve at the originally designed operating point (Fig. 4.12).

Although operationally it may be possible to tolerate these options, they should not be accepted from an energy-efficiency point of view as higher than required flow or pressure results in higher pumping power consumption.

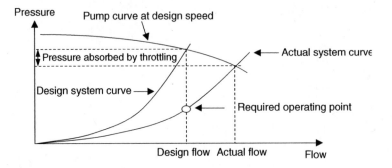

Figure 4.12   Oversized pump with throttling.

In such a situation, the design flow rate can be achieved by reducing the impeller diameter (trimming impeller) or reducing the speed of the pump (using a variable speed drive). Variable speed drives (VSDs) are also sometimes called variable frequency drives (VFDs) or adjustable frequency drives (AFDs). They are devices that can convert the frequency of the utility power supply and provide an adjustable output voltage and frequency to vary the speed of motors.

The resulting energy saving due to speed reduction and impeller diameter reduction is illustrated in Figs. 4.13 and 4.14.

The decision to reduce the speed or impeller diameter will depend on the relative cost for the two options. Normally, reducing the pump speed using a VSD is preferable since it can be used to vary the pump speed and capacity if the load changes in the future. Also, reducing the impeller diameter can result in a bigger drop in pump efficiency when compared to the use of VSDs for reduction of speed. However, if the existing pump is old and due for replacement, the option of replacing the pump with a correctly-sized new pump should also be considered.

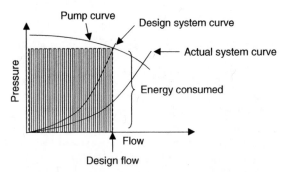

Figure 4.13   Energy consumed by pump that is "throttled" to give the design flow.

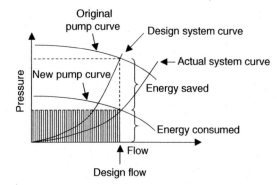

Figure 4.14 Energy consumed by the same pump if impeller diameter or speed is reduced to give the design flow.

Usually, measurements of system flow rate and pressure are required for estimating the savings potential through reducing pump speed, trimming the impeller, or replacing the pump.

The simplest method of estimation is by using the pump affinity laws, as illustrated in Example 4.2. The value of measured flow can be used together with the operating pump speed to find the new speed and thereby the required flow. Thereafter, the computed new speed can be used to estimate the resulting power consumption, as illustrated below.

Example 4.2 A pump is designed to pump 10 L/s of water when operating at 1400 rpm. Under actual operating conditions, the water flow is 15 L/s and the pump motor consumes 15 kW. Calculate the reduction in pump power consumption if the pump speed is reduced to provide the design water flow of 10 L/s.

From pump affinity laws, the pump speed can be reduced to give the design flow as follows:

$$\text{New pump speed} = 1400 \times (10/15) = 933 \text{ rpm}$$
$$\text{New power consumption} = 15 \times (10/15)^3 = 4.5 \text{ kW}$$
$$\text{Reduction in pump power consumption} = 15 - 4.5 = 10.5 \text{ kW}$$

A better, more accurate estimate can be made if the pump curves are available for the particular pump in use. The measured flow rate and pressure, which is the current operating point, can be plotted on the pump curve, as shown in Fig. 4.15. Then, the system curve can be plotted assuming a parabolic relationship. On this system curve, the required operating point can be marked by drawing a vertical line at the value of the required flow. Manufacturers' pump curves can then be used to find the new pump speed or impeller diameter by interpolation, as shown in Fig. 4.15. Once the required impeller size or speed is selected, the manufacturer's curves can also be used to estimate the motor power consumption at the new operating point.

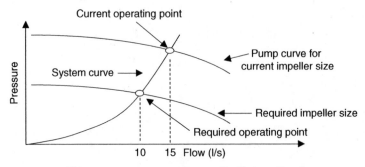

Figure 4.15 Using pump curves to select a suitable impeller size.

4.5.2 Variable flow

The sizing of pumps is done based on a set of design conditions, and in buildings, since their cooling or heating load varies with time, pumps are sized to satisfy peak load conditions.

Water distribution systems use either 2-way or 3-way valves to vary the flow of water through the coils and thereby control the amount of cooling or heating performed by the coils, as shown in Figs. 4.16 and 4.17.

In systems using 2-way valves, water flow through coils is controlled by restricting the flow directly using the valves. In systems using 3-way valves, flow through the coils is controlled by bypassing some of the water from the inlet of the coil to the outlet of the coil.

In systems using 3-way valves, the flow of water through the system has to be kept constant irrespective of the load. Therefore, for example, if the system shown in Fig. 4.17, is designed to operate two chillers at peak load (the third chiller is kept as standby), then two chilled water pumps need to be operated even at times when the load can be satisfied by one chiller. This necessitates operating the second chiller when it is

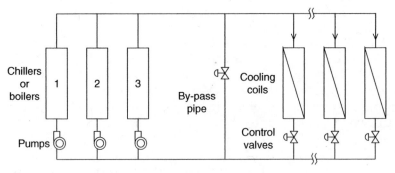

Figure 4.16 Water pumping systems with 2-way valves.

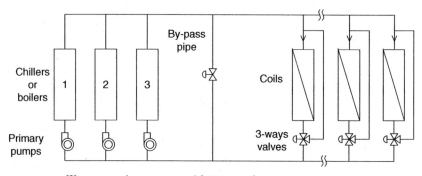

Figure 4.17 Water pumping systems with 3-way valves.

not required or running the second pump when the chiller is not in operation (leading to possible rise in chilled water supply temperature). Owing to these shortcomings, systems using 3-way valves are not popular and those with 2-way valves are generally preferred.

In variable water flow systems with terminal units having 2-way modulating valves, reduction in cooling load causes the modulating valves to close, resulting in reduced water flow. If the pumps are constant speed pumps, this causes the pump operating point to move along the pump performance curve by increasing the system pressure (Fig. 4.18).

However, in such situations, if the speed of the pump can be varied, the pump can be operated at a lower speed to provide the required flow. Since the power consumed by pumps is directly proportional to the cube of the speed (cube law), significant savings can be achieved if pumps are fitted with VSDs (Fig. 4.19).

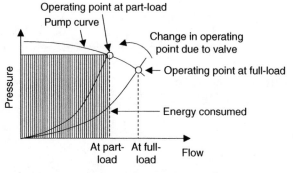

Figure 4.18 Pump operating point for system with 2-way valves and constant speed water pumping (valve control).

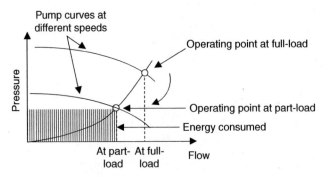

Figure 4.19   Pump operating point for system with 2-way valves and variable speed pumping (pump control).

4.5.3   Primary-secondary systems

Two types of variable speed pumping systems are used for chilled water pumping. The most common system is the primary-secondary pumping system where two sets of pumps are used, as shown in Fig. 4.20. The first set of pumps, or primary pumps, is used to pump water through the chillers. The second set of pumps or secondary pumps, is used to pump water through the building distribution system. The primary pumps operate at constant speed and provide a fixed chilled water flow through each chiller in operation. The secondary pumps are operated based on the chilled water requirements of the building. The speed of the secondary pumps are varied based on the cooling load, which is normally sensed using a differential pressure sensor located at the hydraulically furthest AHU (air handling unit). The primary and secondary loops are hydraulically isolated using a decoupler pipe. The purpose of the decoupler is to ensure that the primary system is not affected by flow

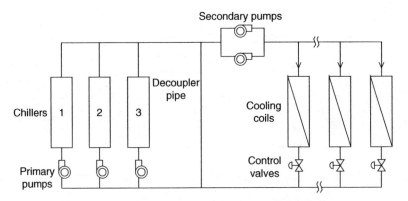

Figure 4.20   Arrangement of primary-secondary pumping systems.

and pressure variations in the secondary system. This enables the primary pumps to maintain a constant flow rate through the chillers regardless of building's cooling load.

The primary pumps circulate a fixed chilled water flow in the primary circuit, while the secondary pumps provide sufficient flow to the cooling coils to satisfy the cooling load. The difference in the chilled water flow between the primary and secondary systems passes through the decoupler pipe. As explained earlier in chiller sequencing, the flow in the decoupler should be from the chilled water supply side to the return side. If the chilled water flow in the decoupler is in the opposite direction, it indicates that there is insufficient flow in the primary circuit and therefore an additional primary pump (together with a chiller) needs to be operated.

Variable chilled water flow systems using primary-secondary pumping are generally preferred since they are able to maintain a constant flow of chilled water through the chillers while varying the flow of chilled water in the distribution system. This satisfies the requirement of many chillers for a constant flow of chilled water through the evaporator tubes to enable stable operation (prevents safety systems shutting down chillers to avoid tube freezing caused by sudden reduction in flow). Constant chilled water flow also helps to maintain the heat transfer coefficient in the evaporator of the chiller.

Although the speed of the primary pumps cannot be reduced at part-load, they are generally small and are sized to pump chilled water only through the primary chiller circuit. This enables the speed of the much larger secondary pumps to be reduced at part load without affecting the performance of the chillers.

Recent improvements in chiller controls have made it possible for chillers to operate with some variation of chilled water flow through the evaporator tubes. This has led to the use of VSDs on pumping systems that have only primary pumps. A typical arrangement of primary pumps with VSDs is shown in Fig. 4.21.

In contrast to primary-secondary systems, this system varies the flow of water through the entire system, including the evaporators of the chillers and the cooling coils. The chillers are sequenced based on the cooling load computed using the flow rate FM and chilled water return and chilled water supply temperature sensors, T_R and T_S, respectively. Motorized valves interlocked to the chillers are used to prevent chilled water circulation through chillers not in operation.

The VSDs of the chilled water pumps are controlled based on the differential pressure sensor DP-2 to maintain a set differential pressure across the furthest AHU. The motorized bypass valve is controlled using differential pressure sensor DP-1 to maintain a minimum differential pressure across the chillers to ensure minimum flow. If the chilled water flow is too low, the differential pressure sensor DP-1 will sense that the

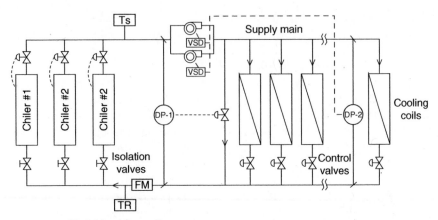

Figure 4.21 Variable primary flow system.

pressure is below set point and will open the bypass valve, enabling some water to by pass and circulate through the chillers. Due to this bypassing of chilled water, the differential pressure sensor DP-2 will sense a drop in pressure and will signal for the speed of the pumps to be increased. The system should be designed to ensure that the flow is maintained within minimum and maximum flow limits for chillers (usually 0.9–3.4 m/s).

One of the main advantages of variable primary flow systems is the cost savings resulting from eliminating secondary distribution pumps and their associated piping.

However, before implementing a variable primary flow system, it is necessary to confirm that the chiller controls are able to support it and that savings achievable is greater than for a primary-secondary system. Example 4.3 illustrates how savings can be estimated for different pumping systems.

Example 4.3 Consider a single building with a 500-RT peak cooling load The pumping power consumed for three options (see Figs. 4.22 to 4.24); constant speed pumping (base case – no VSDs), variable primary flow pumping (VSDs on primary pumps), and primary-secondary pumping (VSD on secondary pumps) are considered.

The first option (Fig. 4.22) with constant speed pumps is where the pumping power remains relatively constant irrespective of the building's cooling load. The system has 2-way valves and the pump operating point "rides" on the pump curve (lower flow–higher pressure) at part load, as explained earlier. The same also applies for systems with 3-way valves, where the flow rate remains constant irrespective of load.

If the chillers are 250 RT capacity each, only two chillers need to be operated at peak cooling load of 500 RT. Therefore, at cooling loads up to 250 RT, only one pump needs to be operated while two pumps are needed at other times.

If the chilled water system is designed for a ΔT (difference in chilled water return and supply temperatures) of 5.6°C (10°F), the chilled water flow rate required for each 250-RT chiller is 37.8 L/s (600 USgpm). Assuming a total system

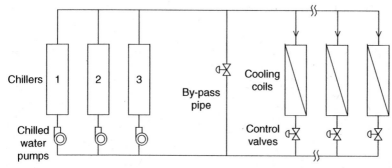

Figure 4.22 Option 1—constant speed pumps.

head of 210 kN/m^2 (70 ft. water), and pump and motor efficiencies are 80 percent each, the theoretical pump power consumption can be calculated as follows:
From Eq. (4.5),

$$\text{Pump brake horsepower} = \frac{\text{flow rate (m}^3\text{/s)} \times \text{pressure difference (N/m}^2)}{1000 \times \text{efficiency}}$$

$$\text{Therefore, P} = \frac{(Q \times \Delta P)}{(1000 \times \eta_p \times \eta_m)} \tag{4.6}$$

where P = pump power consumption, kW
　　Q = flow rate, m^3/s
　　ΔP = pump head, N/m^2
　　η_p = pump efficiency
　　η_m = motor efficiency

For operating two pumps,

$$P = [0.0378 \times 2 \times 210 \times 10^3]/[1000 \times 0.8 \times 0.8] = 24.8 \text{ kW}$$

Therefore, when two chillers operate, the pumping power will be 24.8 kW. When one pump operates, the motor power can be taken as 12.4 kW (in actual operaton, it will not be exactly half).

The second option (Fig. 4.23) considers variable speed primary pumps. The theoretical pumping power is estimated using the affinity laws. Although the actual pump power consumption may not follow this relationship exactly due to drop in pump efficiency at lower speeds, it is convenient to use this relationship for estimation purposes. It is also assumed that the minimum chilled water flow required by the chillers is 50 percent (pump speed can be reduced by 50 percent).

In option 3 (Fig. 4.24), which is a primary-secondary system, the total system head is assumed to be made of 60 kN/^{2} (20 ft. water) for the primary circuit and 150 kN/m^2 (50 ft. water) for the secondary circuit. The pump and motor efficiencies are taken to be 80 percent. Using Eq. (4.6), the motor power consumption for primary pump and secondary pump, at 100 percent speed, can be computed as follows:

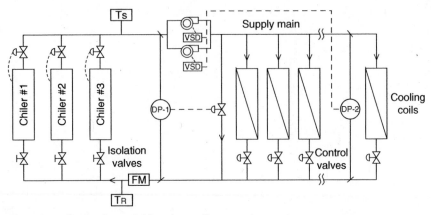

Figure 4.23 Option 2—variable primary flow.

Primary pumps

$$P = [0.0378 \times 2 \times 60 \times 10^3]/[1000 \times 0.8 \times 0.8] = 7.0 \text{ kW (2 pumps)}$$

Secondary pumps (at full capacity)

$$P = [0.0378 \times 2 \times 150 \times 10^3]/[1000 \times 0.8 \times 0.8] = 18.0 \text{ kW (2 pumps)}$$

Based on this, for simplicity, primary pumps and secondary pumps are assumed to consume 3.5 kW and 9 kW each, respectively, at full load. At part load, the secondary pumps are assumed to follow the pump affinity laws for flow, head, and power consumption. The minimum operating speed for the pumps is taken as 40 percent.

The pumping power consumed for the three options for a typical load profile is summarized in Table 4.2.

Notes:

Option 1—Pump power for option 1 is taken as 12.4 kW for one pump and 24.8 kW for two pumps.

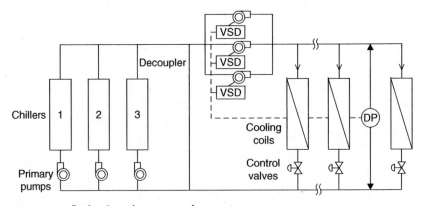

Figure 4.24 Option 3—primary secondary system.

TABLE 4.2 Summary of Pumping Power for Options 1, 2, and 3

Loading	Building cooling load (RT)	Pumping power (kW)		Option 3		Daily operating hours	kWh consumption per day		
		Option 1	Option 2	Primary	Secondary		Option 1	Option 2	Option 3
100%	500	24.8	24.8	7.0	18	3	74.4	74.4	75
90%	450	24.8	18.1	7.0	13	8	198.4	144.6	161.0
80%	400	24.8	12.7	7.0	9	4	99.2	50.8	64.9
70%	350	24.8	8.5	7.0	6	3	74.4	25.5	39.5
60%	300	24.8	5.4	7.0	4	3	74.4	16.1	32.7
50%	250	12.4	12.4	3.5	2	2	24.8	24.8	11.5
40%	200	12.4	6.3	3.5	1	1	12.4	6.3	4.7
						Total	558	343	389

Option 2—Pump power is taken to vary according to the cube law up to a minimum speed of 50 percent.

Option 3—Primary pump power is taken as 3.5 kW each (7 kW for two pumps) and the secondary pump power consumption is taken to vary according to the cube law up to a minimum speed of 40 percent.

Table 4.2 shows that for the particular operating pattern considered (load profile and operating hours), the lowest daily power consumption is achieved for options 2 and 3. Although option 2 has a 12 percent lower power consumption than option 3, option 3 may still be preferred from an operational standpoint.

In a case where a single central chilled water plant supplies chilled water to multiple buildings that have different chilled water requirements (different cooling loads and pressure heads), primary-secondary systems would be more advantageous as the secondary pumps can be sized to suit the requirements of different buildings (see Fig. 4.25).

If buildings A and B are office towers that have a higher chilled water flow and pressure head requirement compared to building C, which is a podium block with retail outlets, the secondary pumps for the podium block can be sized to match the lower flow and head requirements of this building. However, if only primary pumps are used in this situation, the pumps have to be sized to satisfy the higher capacity requirements of the office blocks and will end up pumping chilled water at higher than required pressure to the podium block, wasting much pumping energy.

Since the pumping energy savings depends on factors such as the load profile, building pressure head, and configuration of buildings, the choice of whether the VSDs should be on primary or secondary pumps (to maximize savings) should depend on the actual installation. Nonetheless,

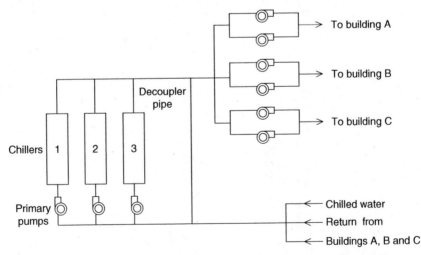

Figure 4.25 Arrangement of a primary-secondary system to serve multiple buildings.

from a practical point of view, it may be better to have a primary-secondary system with VSDs on secondary pumps since it would enable higher variations in the flow without affecting the chiller performance. However, it should be noted that when reducing the speed of pumps, the speed should not be reduced by more than about 40 percent to ensure sufficient lubrication of pump seals and cooling of the motors.

4.5.4   Reset of variable flow pump set point

Further energy savings from chilled water pumping can be achieved by varying the differential pressure set point used for controlling the VSD speed according to demand. In such a system, a building automation or energy management system (BAS or EMS) can be used to monitor the position of the control valves at the AHUs and reduce the differential pressure set point, while ensuring that none of the valves are starved of chilled water. A possible control strategy is shown in Fig. 4.26.

The position of all the AHU chilled water modulating valves are monitored and the valve that is open most is determined. The value of valve position is compared with limits set for adjusting the set point. For example, if the maximum valve position is less than 70 percent open, it indicates that the other valves are open even less than 70 percent and therefore the pressure set point can be reduced further. Similarly, when load conditions change, if a valve is open more than 90 percent, the set point will be increased to prevent the cooling coil from being starved of

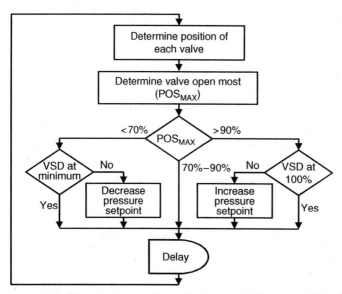

Figure 4.26 Control strategy for optimizing variable speed chilled water pumps.

chilled water. This constant readjustment of the set point, used for controlling the speed of the secondary chilled water pumps, can help to further improve the efficiency of chilled water pumping systems.

4.5.5 Optimizing condenser cooling systems

As in the case of chilled water pumping systems, if the condenser water pumps are oversized for the application and the pumps are either providing too much flow or the flow is throttled using valves, pump capacity can be reduced by lowering the operating pump speed or reducing impeller diameter. Both measures will normally result in significant savings in energy consumption.

The choice of whether to reduce the impeller diameter or reduce the pump speed will depend on the cost for each option. Normally, reduction in pump speed using a VSD is preferred due to flexibility in implementation. This option also results in better operating efficiency as reducing the impeller diameter can significantly effect pump efficiency.

Condenser water pumping using variable flow systems (where pump speed is varied based on the load) does not normally yield significant energy savings. This is because at part load, full design condenser water flow results in a lower condenser water return temperature (when load drops, if flow rate is the same, the ΔT drops, and because the condenser water supply temperature is the same, the condenser water return temperature is reduced), which results in better chiller efficiency. As shown in Fig. 4.27, condenser water enters the condenser at temperature T_S and leaves the condenser at temperature T_R. The condensing temperature, T_C, depends on the condenser approach temperature, which is the difference between the condensing temperature and condenser water leaving temperature $(T_C - T_R)$. The condenser approach temperature depends on heat transfer characteristics such as the thickness of the condenser tubes and fluid velocity in the tubes. Therefore, the approach temperature remains constant at part load, and lower condenser water return temperature (T_R) results in lower condensing temperature (T_C) at part load.

Lower condensing temperature and pressure leads to lower compressor power, as seen in the p-h diagram of an ideal refrigeration cycle (Fig. 4.28). The energy savings made due to improvement in chiller efficiency at part load normally exceeds the possible pumping savings that can be achieved by reducing condenser water flow rate at part load.

4.5.6 Pressure drop (ΔP) across chillers

When water flows through the evaporator or condenser tubes of chillers, a pressure drop results due to the flow resistance offered by

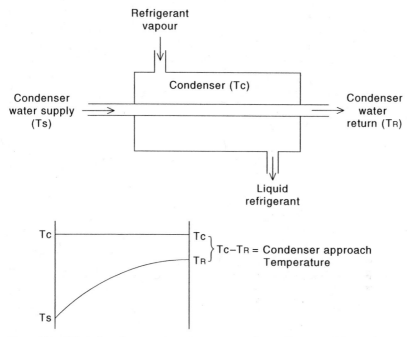

Figure 4.27 Effect of condenser water temperature on the performance of the condenser.

the tubes. The pressure drop is dependent on the flow velocity in the tubes, which in turn is dependent on the water flow rate and tube cross-sectional area.

Higher pressure drop across a chiller evaporator and condenser necessitates the chilled water and condenser water pumps to work against a higher pressure difference to deliver the same flow rate. This results in

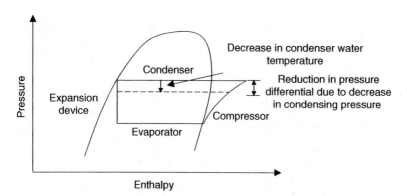

Figure 4.28 Effect of reducing condenser water temperature.

higher power consumption by the pump, since pumping power is directly proportional to the flow and pressure head (kW ∝ Q × ΔP).

Usually, a pressure drop across chillers is not considered an important criteria when selecting chillers as more emphasis is given to criteria such as efficiency, cost, and brand of chiller. This some times leads to selection of chillers with high pressure drops, which, in turn, leads to higher energy consumption by the pumps, as shown in Example 4.4.

Example 4.4   Consider the following two chiller options A and B.

	Option A	Option B
Chiller capacity	500 RT	500 RT
Pressure drop across	90 kN/m^2	30 kN/m^2
evaporator	(30 ft water)	(10 ft water)
Chilled water flow rate	75.6 L/s	75.6 L/s
	(1200 USgpm)	(1200 USgpm)

Using the following equation, the theoretical pump power consumption to overcome the resistance across the evaporator can be calculated as follows (assuming efficiency of 100 percent):

$$\text{Pump kW} = [\text{Flow in m}^3/\text{s} \times \text{Pressure difference in N/m}^2]/1000$$

$$\text{Option A, pump kW} = [0.0756 \times 90{,}000]/1000 = 6.8 \text{ kW}$$

$$\text{Option B, pump kW} = [0.0756 \times 30{,}000]/1000 = 2.3 \text{ kW}$$

$$\text{Pump kW savings due to Option B} = (6.8 - 2.3) = 4.5 \text{ kW}$$

The same exercise can be performed to compute the extra pumping power consumption due to different condensers' pressure drops for chillers.

4.5.7   Pressure losses in pipes and fittings

When a liquid flows through a piping system, head or pressure losses take place due to fluid friction in the piping and resistance offered to the flow by the various devices such as valves, strainers, and bends used in the piping system.

The friction losses depend on the pipe material, length of the piping, fluid velocity, and properties of the fluid. Therefore, for a given fluid such as water, the friction losses can be reduced by minimizing pipe length and reducing flow velocity (increasing pipe diameter).

On the other hand, for a particular flow velocity, losses due to various fittings and devices installed on piping systems depend on the design of the device or fitting. Therefore, for a given flow velocity, different types of valves can have different losses associated with them.

The losses for different types of valves can be significantly different and, therefore, one has to be careful when selecting valves for different applications. For instance, globe type valves are commonly used as isolation valves in chilled water and condenser water piping systems. These valves have a high-pressure drop even when they are fully open due to the change in direction the flow has to make when passing through them. On the other hand, butterfly valves, when fully open, offer little or no resistance to the flow. Therefore, to minimize pumping energy consumption, valves with low resistance (when fully open), such as butterfly valves, should be used for flow isolation.

Further, pipe fittings such as bends, elbows, tees, and flow transition devices should also be selected to minimize head losses in the system.

4.5.8 Condenser water systems for package units

Some buildings use water-cooled package units to provide air-conditioning. In such systems, the package units are installed in different parts of the building to serve the different areas. Water-cooled package units reject the heat absorbed to the condenser water in the condenser of each unit. Such package units are normally served by a central condenser water system consisting of pumps, cooling towers, and a network of piping, as shown in Fig. 4.29.

Very often, in buildings using such systems, individual package units are scheduled to operate at different times, depending on the requirement of each user. This can result in condenser water circulating through package units that are not in operation if individual package units cannot be isolated from the condenser water system. This leads to pumping of more condenser water than necessary and wastage of pumping power.

Such situations can be avoided if on-off valves are installed on the condenser water pipes serving each water-cooled package unit and

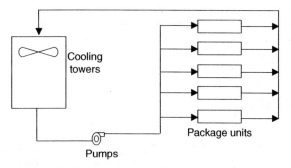

Figure 4.29 Arrangement of a condenser water system serving package units.

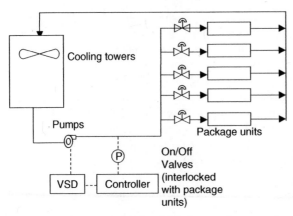

Figure 4.30 Arrangement of variable flow condenser water system.

are linked to the operation of the package units. This will enable the condenser water flow to each unit to be automatically switched on or off, depending on its operation. The condenser water pumps can be fitted with variable speed drives and a control system to maintain a minimum set pressure (P) for the condenser water system, as shown in Fig. 4.30.

4.5.9 Efficiency of pumps

The equation for pump brake horsepower (discussed earlier) is:

Pump brake horsepower

$$= \frac{\text{flow rate } (m^3/s) \ \times \ \text{pressure difference } (N/m^2)}{1000 \ \times \ \text{efficiency}}$$

The overall pumping efficiency refers to, both, the pump and motor efficiencies, and pumping power can be expressed as follows:

$$P = \frac{(Q \times \Delta P)}{(1000 \times \eta_p \times \eta_m)}$$

where P = pump power consumption, kW
  Q = flow rate, m^3/s
  ΔP = pump head, N/m^2
  η_p = pump efficiency
  η_m = motor efficiency

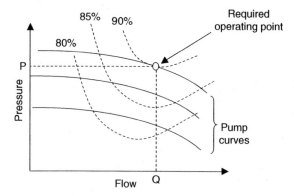

Figure 4.31   Operating point for pump A.

Therefore, to minimize pumping power, the pump efficiency should be as high as possible. This can be done be selecting suitable pumps, which have a high efficiency (above 85 percent) at the desired operating point.

Figures 4.31 and 4.32 show two sets of pump curves for pump A and pump B. As the figures show, both pumps are able to operate at the desired operating point of flow rate, Q, and pressure, P. However, based on the performance curves, pump A will operate at 90 percent while pump B will only be able to operate at 78 percent (estimated by interpolation) at the desired operating point.

Pump efficiency also depends on capacity, and bigger capacity pumps tend to be more efficient than smaller capacity ones. Therefore, it is generally better to have a few higher capacity pumps than many smaller capacity pumps to perform the same duty. This is the advantage of having variable speed chilled water pumping systems where

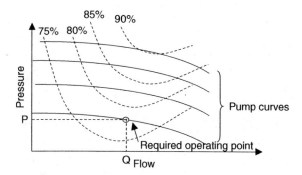

Figure 4.32   Operating point for pump B.

the number of pumps does not have to be the same as the number of chillers because pump capacity can be varied to match load requirements (Fig. 4.23).

Example 4.5 A pumping system requires a pump with a capacity of 80 L/s and a head of 150 kN/m². Calculate the savings that can be achieved in pump power if a pump having 90 percent efficiency is used instead of a pump with 78 percent efficiency for this application.

From Eq. (4.6),

$$P = \frac{(Q \times \Delta P)}{(1000 \times \eta_p \times \eta_m)}$$

If the motor efficiency η_m is taken as 1.0,
Saving in pump power

$$= \frac{(Q \times \Delta P)}{1000}\left(\frac{1}{\eta_{78\,\%}} - \frac{1}{\eta_{90\%}}\right)$$

$$= \frac{(0.08 \times 150 \times 10^3)}{1000}\left(\frac{1}{0.78} - \frac{1}{0.9}\right)$$

$$= 2.05 \text{ kW}$$

4.6 Summary

Pumps are used in buildings mainly for air-conditioning and heating systems. Pumping systems normally account for the second highest energy consumption in buildings. This chapter covered the basics of pumping systems, such as open and closed systems, pump characteristics, and pump curves and system curves, followed by an overview of the pumping systems used in buildings for air-conditioning and heating systems. Thereafter, various energy management strategies for building chilled water, hot water and condenser water pumping systems were described.

Review Questions

4.1. A pump delivers 200 L/s at 1200 rpm and consumes 45 kW. If the pump speed is reduced to 1100 rpm, what will the new flow rate and power consumption be?

4.2. A pump is selected to provide 20 L/s of water when operating at 1200 rpm. Under actual operating conditions, the water flow is 35 L/s and the pump motor consumes 25 kW. What will the reduction in pump power consumption be if the pump speed is reduced to provide the design water flow of 20 L/s?

4.3. The condensers of two different chillers of equal capacity have pressure drops of 80 kN/m^2 and 40 kN/m^2, respectively. If the water flow rate required is 150 L/s, calculate the saving in pump power for the condenser water pump if the chiller with the lower pressure drop is used instead of the chiller with the higher pressure drop.

4.4. A pumping system requires a pump with a capacity of 40 L/s and a head of 120 kN/m^2. If a pump having 85 percent efficiency is used for this application instead of a pump with 65 percent efficiency, what is the saving in power required to drive the pump?

Cooling Towers

5.1 Introduction

Cooling towers are used to reject heat from air-conditioning systems and process-cooling systems. They reject heat into the atmosphere through sensible and latent heat transfer. In air-conditioning systems, cooling towers are used with water-cooled central chiller systems and water-cooled package units. A typical application of cooling towers in central water-cooled chiller systems is shown in Fig. 5.1.

Cooling towers generally consist of water spray systems, "fill" packing material, and fans. Spray systems are used to spread the warm water being cooled over the fill packing, which acts as a heat transfer medium by increasing the contact surface area. Fans are used to induce ambient air flow through cooling towers to facilitate heat transfer between the warm water and the ambient air.

Cooling towers reject heat mainly by evaporative cooling. When water is sprayed in cooling towers, some of the water evaporates, absorbing heat from the surrounding water, thereby cooling it. The amount of latent heat transferred depends on the moisture content of the air; the more dry the air (lower the wet-bulb temperature), the more latent heat that will be transferred. In addition, sensible cooling also takes place between the warmer water and colder air. The amount of sensible cooling depends on the dry-bulb temperature of air. Therefore, the amount of heat rejected from cooling towers depends on, both, the dry-bulb and wet-bulb temperatures of the outdoor air.

Cooling towers used for building applications are normally of the counterflow or crossflow type. Counterflow cooling towers have square or round cross sections and take air from the sides. Crossflow cooling

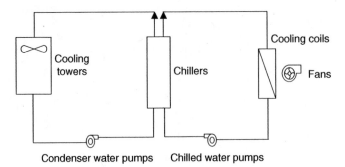

Figure 5.1 Central chiller system with cooling towers.

towers have rectangular cross sections and take air only from the two long sides. The air flow in cooling towers can be either induced draft, where fans pull air out of cooling towers, or forced draft, where fans force air into cooling towers. Cooling towers can be forced-draft counterflow (Fig. 5.2), induced-draft counterflow (Fig. 5.3), forced-draft crossflow (Fig. 5.4), or induced-draft crossflow (Fig. 5.5).

Cooling towers are manufactured in different sizes and capacities to match various cooling requirements. They are selected for a particular application based on the entering temperature of the warm water to be cooled, leaving temperature to which the water needs to be cooled, the flow rate of the water, and the wet-bulb and dry-bulb temperatures of the ambient air.

Cooling towers used in central air-conditioning systems are commonly sized to receive water at 35°C (95°F) and provide water to the chillers

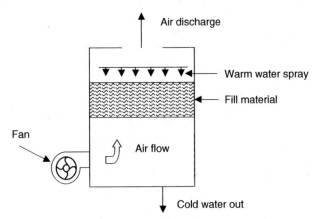

Figure 5.2 Arrangement of forced-draft counterflow tower.

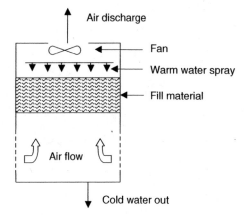

Figure 5.3   Arrangement of induced-draft counterflow cooling tower.

at 29.4°C (85°F). This difference between return and supply temperatures of water is called the "range." Since cooling tower performance depends on the wet-bulb temperature of air, they are normally selected to operate at a specific "approach" temperature, which is the difference between the temperature of water leaving the cooling tower and the wet-bulb temperature of the air.

Theoretically, cooling towers are able to cool water to the wet-bulb temperature of the air. However, this would require a very large cooling tower surface area. Therefore, cooling towers are normally designed to economically cool water to an approach temperature of about 2.8°C (5°F).

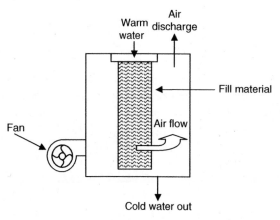

Figure 5.4   Arrangement of forced- draft crossflow cooling tower.

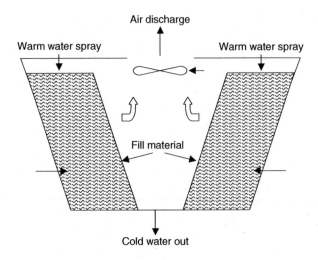

Figure 5.5 Arrangement of induced-draft crossflow cooling tower.

5.2   Energy Saving Measures for Cooling Towers

5.2.1   Cooling tower sizing

Cooling towers in central air-conditioning systems need to reject not only the heat removed from the air-conditioned space but also the heat added by the compressors of chillers. The heat of compression added by a compressor depends on the efficiency of the chiller, and is usually about 25 percent of the cooling load. Therefore, cooling towers need to be sized to reject approximately 125 percent of the chiller cooling capacity.

Sizing of cooling towers is important to ensure that chiller efficiency is optimized. If the cooling towers are undersized for the chillers, the condenser water supply temperature will rise and chiller efficiency will be reduced. Similarly, if the condenser water temperature from the cooling towers can be lowered, chiller efficiency can be improved.

Cooling towers are theoretically able to produce condenser water at the wet-bulb temperature of air. However, to achieve this, it is necessary to have cooling towers with higher surface area and air flow. Although such a design will lead to better chiller efficiency (due to lower condenser water temperature), it will also result in higher capital cost and operating cost as extra cooling towers need to be used.

Also, for the same cooling load, if the cooling tower surface area is increased (oversized cooling tower is used), the air flow can be reduced, leading to lower cooling tower fan power but higher first cost for the cooling towers. Therefore, the sizing of cooling towers is a compromise between the initial capital cost of the cooling towers and the running cost of the chillers and cooling towers.

The selection of cooling towers should be based on the condenser water supply and return temperature requirements of the chiller and the design conditions for outdoor wet-bulb temperature for the region. The most common design conditions are condenser water return from chillers at 35°C (95°F), condenser water supply to the chiller at 29.4°C (85°F), and wet-bulb temperature of 26.7°C (80°F). Often cooling tower suppliers rate cooling tower capacities at different operating conditions. If the cooling tower capacity is rated at a wet-bulb temperature higher than what the cooling tower is going to experience, the resulting capacity of the cooling tower will be lower than the rated value during operation. Therefore, in such situations, the rated capacity of the cooling tower needs to be "derated" to account for different operating conditions. As a rule of thumb, cooling towers with "nominal capacity" (cooling tower rated capacity) of 1.5 times the chiller's maximum rated capacity are used.

5.2.2 Capacity control

The operation of cooling towers is usually interlocked with that of chillers so that when chillers are switched on or off, based on the cooling load, the operation of cooling towers also follows suit. In climates with low ambient temperatures, condenser water supplied to the chillers has to be maintained above the minimum temperature required by chillers.

In such situations, a bypass valve is used in the condenser water circuit to enable some condenser water to bypass the cooling towers, as shown in Fig. 5.6. A control system is used to sense the temperature of condenser water entering the chillers and maintain it by varying the flow of condenser water to the cooling towers. However, such control systems are not required in climates where ambient conditions do not result in low condenser water temperature.

The capacity of cooling towers is dependent on the air flow through them. Therefore, when the chillers operate at part load (when cooling load is low), the amount of heat to be rejected at the cooling towers is also low. Under such operating conditions it is not necessary to run the cooling towers at full capacity. The cooling tower capacity can be reduced by reducing the air flow, which would result in lower fan energy consumption.

One way of achieving this is by cooling tower fan cycling, where some fans are switched on or off to control condenser water temperature. This, however, can result in a swing in condenser water temperature and can cause premature wear and tear of motor drives.

A better way is to use variable speed drives (VSDs) to control cooling tower fan speed. As shown in Fig. 5.7, the speed of the cooling tower fans in operation can be modulated to maintain a set temperature.

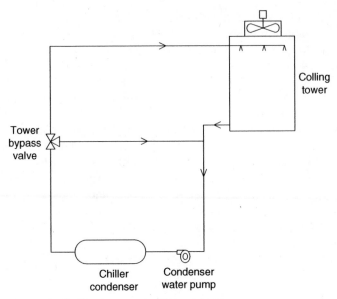

Figure 5.6   Cooling tower water flow control.

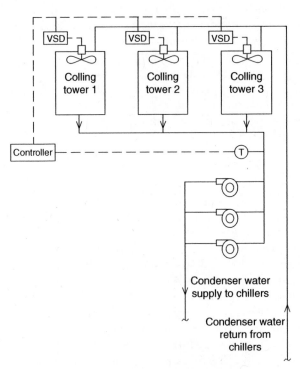

Figure 5.7   Cooling towers with variable speed fans.

The easiest control strategy is to maintain the condenser water supply temperature at the design value. Therefore, for a system designed to provide condenser water at 29.4°C (85°F), the control system can be used to provide condenser water at this set value by modulating the cooling tower fan speed. However, it should be noted that in cooling towers where the warm water distribution system is driven by the fan motor, reducing fan speed can lead to poor water distribution. In such cases, if VSDs are used to control fan speed, it is recommended to set a minimum speed.

Since, theoretically, the power consumed by fans is proportional to the cube of the fan speed, when the cooling load drops to 80 percent, the speed of the cooling tower fans can also be reduced accordingly, resulting in a drop of about 50 percent $(0.8^3 = 0.51)$ in power consumption. Therefore, this control strategy can lead to very significant savings from cooling tower fans at part load. Use of VSDs to control cooling tower capacity, rather than fan staging, also leads to reduction in the wear and tear of the drives due to lower fan speed and less drift losses (water losses) due to lower air velocity.

Example 5.1 Consider a cooling tower of capacity 500 RT that has a 15 kW fan (constant speed) and is designed to cool water from 35 to 30°C.

If, under normal operating conditions, the temperature of the warm water entering the cooling tower is at 32°C, this indicates that the load on the cooling tower is only 40 percent (2°C divided by 5°C).

Therefore, if a variable speed drive is installed on this cooling tower, the fan will operate at approximately 40 percent of its speed to maintain the required leaving water temperature of 30°C.

The theoretical fan power consumption will be $(0.4)^3 \times 15 = 1$ kW

Saving in power consumption will be (15 − 1) = 14 kW

Figure 5.8 shows the relationship between cooling tower efficiency in kW/RT (fan power divided by amount of heat rejection) for three possible fan operating strategies—constant speed fan (without fan cycling), fan cycling, and using VSD.

Cooling tower performance also depends on the ambient wet-bulb temperature. When the wet-bulb temperature drops from the design value during nighttime or during different seasons of the year, the cooling towers are able to economically provide water at a lower temperature, while maintaining a fixed approach temperature. Since chiller efficiency depends on condenser water temperature, this can lead to better efficiency of the chillers. Therefore, a further enhanced control strategy can be used for cooling tower fan speed control, as shown in Fig. 5.9. In this system, the cooling tower approach temperature (difference in condenser water supply and wet-bulb temperature) can be used instead of condenser water supply temperature to control the tower fan speed and further optimize the cooling tower and chiller energy consumption.

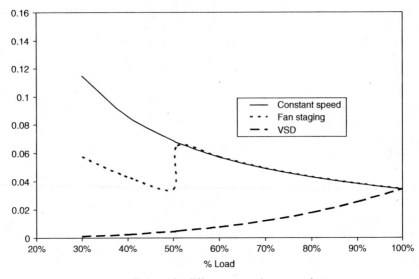

Figure 5.8 Cooling tower efficiency for different operating strategies.

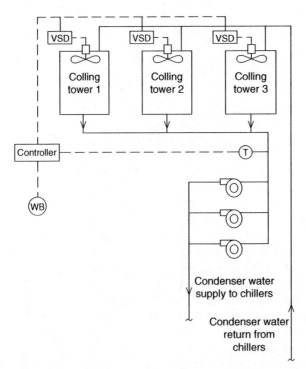

Figure 5.9 Cooling tower fan control to optimize condenser water temperature.

Since cooling tower capacity depends on air flow, operating two identical towers at half the fan speed will result in capacity that is equivalent to operating one tower at full speed. Since, theoretically, the power consumption of the fans is proportional to the cube of their speed; the power consumption of each tower fan will drop to 12.5 percent ($0.5^3 = 0.125$) at half the speed. This will result in a combined power consumption of 25 percent (12.5 percent $\times$ 2), which is a saving of 75 percent of the fan power consumed if only one cooling tower is operated. Therefore, operating more cooling towers in parallel to meet the same load will allow lower fan speed, leading to lower fan power consumption. An additional benefit of running more cooling towers is that lower pumping power is required due to lower pressure losses across the cooling tower spray nozzles (due to lower water flow through each tower).

Almost all installations have one or more extra cooling towers as standby to allow servicing and repair of running cooling towers. Therefore, if the standby towers are run in parallel with the duty cooling towers and if all cooling tower fans are fitted with VSDs, they can be run at lower speed to produce the same amount of heat dissipation.

Example 5.2 A particular installation has three cooling towers (each of 22 kW), of which only two cooling towers are operated and the third cooling tower acts as a standby tower. If all three cooling towers can be run at two-third capacity each (three towers running at two-third capacity will provide the equivalent capacity of two towers), the saving made in fan power can be estimated as follows:

The fan power consumption of each tower will be $22 \times (2/3)^3 = 6.5$ kW (30 percent of full load power).

The total power consumption of all three fans will be $6.5 \times 3 = 19.5$ kW.

Therefore, total savings will be $(44 - 19.5) = 24.5$ kW.

The above savings estimation is based on the system operating at the full load capacity of two cooling towers. Further savings will also be achieved at part-load conditions, as shown in Table 5.1.

TABLE 5.1 Energy Savings from Cooling Towers Fitted with Variable Speed Drives

Cooling load (RT)	Without VSDs		With VSDs				
	No. of towers	Total fan power (kW)	No. of towers	Fan speed (%)	Total fan power (kW)	Operating hours (%)	Savings/day (kWh)
1000	2	44	3	0.67	19.6	10%	58.7
900	2	44	3	0.60	14.3	30%	214.2
800	2	44	3	0.53	10.0	20%	163.1
700	2	44	3	0.47	6.7	10%	89.5
600	2	44	3	0.40	4.2	10%	95.5
500	1	22	2	0.50	5.5	10%	39.6
400	1	22	2	0.40	4.4	5%	21.1
300	1	22	2	0.40	4.4	5%	21.1
						Total	702.8

In Table 5.1, the fan power consumption to satisfy a particular cooling load is computed for the two cases; with and without VSDs. In case of using VSDs, it is assumed that additional cooling towers are operated to enable operation of fans at lower speed and that the minimum speed of the cooling tower fans is set at 40 percent.

For example, when the load is 800 RT, if the fans are equipped with VSDs, they can be operated at 80 percent of the speed (800/1000 = 80 percent). Now, if an additional cooling tower can be operated, this load can be shared by three cooling towers instead of two (each running at two-thirds capacity) and the fan speed will be (80 percent × 2/3 = 53.3 percent). The fan kW is computed by multiplying the cube of speed by number of fans and fan kW at full speed (0.5333 × 22 kW × 3 towers = 10 kW). In this estimation, it is assumed that the efficiency of the fan, motor, and VSD remain constant.

This computation shows that energy savings of 702.8 kWh a day can be achieved for this installation if the cooling tower fans are fitted with VSDs.

5.2.3   Condenser water reset

As explained earlier, in Chapter 2, the operating efficiency of chillers can also be improved by reducing the condenser water temperature. An improvement of 1 to 2 percent can be achieved in chiller efficiency by reducing the condenser water temperature by 0.6°C (1°F) due to the reduction in pressure differential across which the compressor has to work (Fig. 5.10).

Cooling towers are designed to cool condenser water to within a few degrees of the wet-bulb temperature. The most common cooling tower design is for a "cooling tower approach temperature" (temperature difference between condenser water supply temperature and wet-bulb temperature) of 2.8°C (5°F). However, under some operating conditions, such as at part load, cooling towers are able to supply condenser water at lower than the designed approach temperature. Since cooling towers are designed to operate at a certain approach temperature, operating them at a lower than the design value can result in reduction of tower efficiency.

Therefore, in installations where the condenser water temperature is not controlled, the condenser water temperature will drop under such

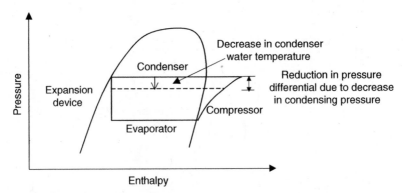

Figure 5.10   Effect of reducing condenser water temperature on chiller efficiency.

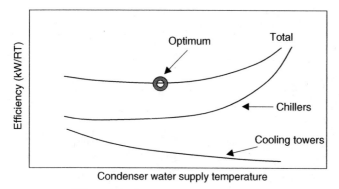

Figure 5.11 Effect of condenser water temperature on chiller and cooling tower efficiency.

off-design conditions, leading to improvement in chiller efficiency. However, this may not be the optimum operating point since the improvement in chiller efficiency may be at the expense of reduced cooling tower efficiency.

Figure 5.11 shows the variation in chiller and cooling tower efficiencies with condenser water temperature. Generally, lower condenser water temperature results in improved chiller efficiency and a drop in cooling tower efficiency (due to the need for working the towers "harder" or operating extra towers to produce condenser water at a lower temperature). Therefore, the overall optimum operating point for a system may not be at the lowest condenser water temperature.

This optimum operating point will vary from one site to another, depending on system design, equipment efficiencies, and configuration. If accurate performance data is not conveniently available for a system, it may be easier to use the control strategy described earlier, where variable speed drives on cooling tower fans are used to maintain a fixed approach temperature.

5.2.4 Condenser water flow rate

Another important aspect of condenser water systems is the water flow rate. As explained earlier, it is essential to provide sufficient condenser water flow to ensure that the heat rejected in the condenser can be dissipated outdoors. The most common design is for a condenser water "range" (difference between condenser water return and supply temperatures) of 5.6°C (10°F). At this design condition, it is necessary to provide 0.19 L/s per RT of cooling capacity (3 USgpm per RT). This required condenser water flow rate is 1.25 times the flow required for chilled water (0.15 L/s or 2.4 USgpm per RT) to account for the heat of compression added by the compressor, which is approximately 25 percent of the cooling load. If sufficient flow is not provided, it will lead to higher condenser water return temperature, leading to higher condensing pressure and, therefore, higher compressor power.

It is better to provide the designed condenser water flow rate even at part load since this results in lower condenser water return temperature and, therefore, better chiller efficiency. However, it has to be ensured that the designed condenser water flow rate is not exceeded. Firstly, it will ensure that pumping energy is not wasted, and secondly, it will ensure that the cooling towers are able to perform satisfactorily. If the cooling towers are supplied with higher than the designed flow rate (sometimes this happens when water flow to the cooling towers is not balanced and some towers get more flow than others), the cooling towers will be unable to provide water at the designed temperature. This will lead to higher condenser water temperature and result in lower chiller efficiency.

Hence, it is essential to have the correct condenser water flow. If the flow is insufficient, the condenser water return temperature will rise, leading to lower chiller efficiency. Similarly, if the condenser water flow is more than that specified for the cooling towers, it will lead to a higher condenser water supply temperature, which too will result in a drop in chiller efficiency.

5.2.5   Installation of cooling towers

Since the performance of cooling towers depend on the air flow through them, cooling towers should be installed such that air can flow freely into them. As shown in Fig. 5.12, cooling towers should be sufficiently spaced so that the air intakes of cooling towers are not too close to each other. Further, cooling tower air intakes should not be less than a minimum distance from obstructions such as walls. The minimum distance to be maintained between cooling towers and from obstructions are normally specified by the cooling tower suppliers, based on the design of their towers.

Care should also be taken to ensure that the warm and moist air being discharged from cooling towers is not recirculated back into the

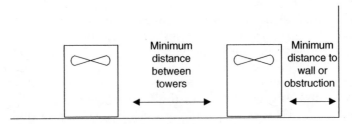

Figure 5.12 Spacing of cooling towers.

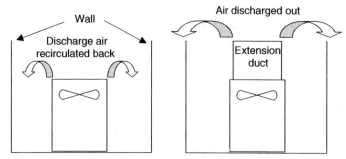

Figure 5.13   Extension duct used to prevent recirculation.

air intakes as this will lead to a drop in tower performance. In some situations, an extension duct can be fitted to the cooling tower discharge to direct the air flow away from the air intakes, as shown in Fig. 5.13.

5.2.6   Condition of cooling tower

The ability of a cooling tower to provide its designed cooling capacity depends on the operation of the water spray system, the fill, and the fan. Therefore, regular maintenance should ensure that the water spray system is able to properly spread the water flow, the fill is in good condition, and the fan is able to operate at the required speed.

If the spray system is defective, warm water may flow directly into the discharge basin of the cooling tower rather than being sprayed onto the fill material that is meant to facilitate the heat transfer between the warm water and the ambient air. Similarly, if the infill is damaged or blocked, water will flow directly into the discharge basin rather than flowing through it, as shown in Fig. 5.14.

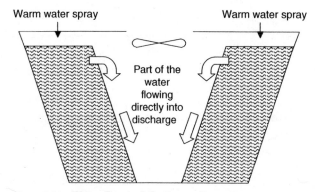

Figure 5.14   Water flow in defective cooling towers.

Since many cooling tower fans use belt drive systems, the tension of the belts should also be checked to prevent belt slippage, which would result in lower fan speed and a drop in cooling tower performance.

All the above aspects, such as ineffective spray systems, damaged fill material, and low fan speed can lead to poor cooling tower performance, which normally results in cooling towers not being able to provide water at the designed temperature. In chiller systems, this will lead to lower chiller efficiency and higher energy consumption.

5.2.7   Water treatment

Another aspect of chiller efficiency, related to cooling towers, is the condition of the condenser water. Since condenser water systems are open systems that require makeup water (to compensate for water evaporated in the cooling towers), water treatment is important to maintain chiller efficiency. Water treatment usually consists of chemical or nonchemical treatment to prevent scaling, corrosion, and fouling of the chiller's condenser tubes. If water treatment is not adequate, fouling and scaling of chiller condenser tubes leads to lowering of chiller efficiency, as explained in Chapter 2.

5.2.8   Free cooling

Vapor migration.   Cooling towers and condenser water pumps can be used to provide "free cooling" in temperate climates during certain seasons, as seen in Chapter 2. One way to achieve this is by switching off the chiller compressor and allowing vapor to migrate from the evaporator to the condenser when the condenser water is at a sufficiently low temperature to allow condensation of the vapor without having to raise its pressure.

Using a heat exchanger.   Another approach is to use water from the cooling towers to either precool or completely cool the return chilled water when the outdoor weather conditions are favorable, as shown in Fig. 5.15. In this system, whenever the condenser water temperature is less than the chilled water temperature, the cooling tower water is used to cool the return chilled water with the help of a heat exchanger. Depending on outdoor conditions and condenser water temperature, the chiller can be turned off completely and the heat exchanger can be used to completely cool the chilled water, or the chiller can be operated at reduced load by using the heat exchanger to precool the return water.

Switching off cooling tower fans.   In some installations, which have a large number of cooling towers to meet daytime peak cooling loads, if

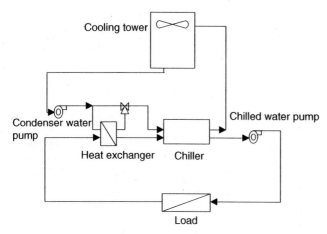

Figure 5.15 Arrangement of free cooling using a cooling tower.

the off-peak (nighttime) cooling load is very low and only a few cooling towers are operated at such times, it may be possible to achieve the required heat rejection by switching off all the cooling tower fans and circulating water through them. This is possible because, although the capacity of cooling towers is proportional to the air flow, they can provide some cooling even when the fans are switched off. Figure 5.16 shows a typical relationship between fan air flow and cooling tower capacity.

As shown in Fig. 5.16, a typical cooling tower can provide about 5 percent of the design capacity when the tower fan is switched off. Therefore,

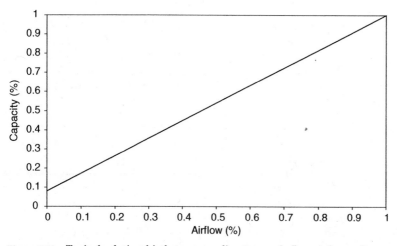

Figure 5.16 Typical relationship between cooling tower air flow and capacity.

if an installation has five cooling towers of 500-RT heat rejection capacity each, they will be able to provide about 125 RT of free cooling (5 × 500 × 0.05 = 125).

5.3 Summary

Cooling towers are used in buildings mainly for providing the heat rejection requirements of air-conditioning systems. Although cooling towers themselves do not consume much energy in building applications, their performance is important as they can have a significant impact on the operating efficiency of air-conditioning chillers. The chapter provided an overview of cooling towers and their application in buildings followed by a number of energy saving strategies to ensure optimum efficiency of air-conditioning systems in buildings.

Review Questions

5.1. A cooling tower is designed to cool 50 L/s of water from 35°C to 30°C at an ambient wet-bulb temperature of 28.5°C.
However, under the following actual operating conditions, the temperature of water leaving the cooling tower is 32°C. What are the possible reasons for the higher than designed cooling tower water temperature and what remedial action can be taken for each of the possible causes?

Actual operating conditions are:

Water flow rate = 48 L/s

Entering water temperature = 34°C

Wet-bulb temperature of air entering the cooling tower = 28°C

5.2. A cooling tower is unable to provide the rated heat rejection capacity when operating at the designed water flow rate, entering water temperature, and ambient wet-bulb temperature. However, it was observed that the wet-bulb and dry-bulb temperatures of air entering the cooling tower were much higher than the ambient wet-bulb and dry-bulb temperatures. Why is the cooling tower not able to provide its rated capacity?

5.3. A cooling tower of capacity 250 RT uses a constant speed fan that consumes 15 kW. If, under actual operating conditions, the load on the cooling tower is only 60 percent of its rated capacity, estimate the savings that can be achieved in fan power if a variable speed drive is installed to operate the fan at 60 percent of its rated speed.

5.4. A particular installation has two cooling towers (each of 15 kW), of which only one cooling tower needs to be operated and the remaining one acts as a standby. Estimate the fan power savings that can be achieved if variable speed drives are installed on both cooling towers and both towers are operated at 50 percent capacity to provide the approximate heat rejection capacity of one cooling tower.

5.5. A cooling tower installation has three 500 RT and two 200 RT capacity cooling towers. If these cooling towers are able to provide up to 5 percent of their capacity when the fans are switched off (free cooling), estimate the maximum total heat rejection capacity that that these cooling towers are able to provide under free cooling conditions with the fans switched off.

6

Air Handling and Distribution Systems

6.1 Introduction

In buildings that are centrally cooled or heated, air is normally treated in air handling units (AHUs) to control moisture content and temperature. Once the air is treated, it is transported and distributed to various parts of the building. A typical air distribution system consists of fans, ducting, dampers, filters, air inlets, and air outlets, as shown schematically in Fig. 6.1.

In such systems, a mixture of outdoor air (provided for ventilation) and part of the air returning from conditioned spaces (return air) is filtered and then treated by the coils. Thereafter, the fan transports the treated air through the supply ducting system, which distributes it in required quantities to the spaces to be conditioned via outlets and dampers. After performing the necessary cooling or heating, air is later returned from the conditioned spaces through the inlets and return ducting. Some of the return air is then recirculated back to the AHU while the balance is expelled to allow sufficient fresh air to be added to the system.

In an air distribution system, the fan provides the necessary energy to move the air by overcoming frictional losses in the ducting and pressure losses due to components in the system, such as filters, coils, and various fittings. The electrical energy required to operate the system can be minimized if the system design is optimized to reduce these losses.

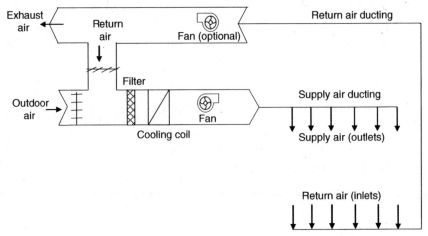

Figure 6.1 Typical air distribution system.

6.2 System Losses

When airflows in ducting systems, there is pressure drop due to *friction losses* and *dynamic losses* caused by change of direction or velocity in ducts and fittings.

Friction losses are due to fluid viscous effects and can be expressed by means of D'Arcy's equation:

$$\Delta P_f = \frac{f \cdot L \cdot v^2}{2 \cdot g \cdot D_m} \tag{6.1}$$

where ΔP_f = frictional pressure drop
$\quad$ f = friction factor
$\quad$ L = length of duct
$\quad$ v = mean duct velocity
$\quad$ g = acceleration due to gravity
$\quad D_m$ = hydraulic mean diameter = $\dfrac{\text{cross-sectional area}}{\text{perimeter}}$

Dynamic losses occur due to change in flow direction caused by fittings such as elbows, bends, and tees and changes in area or velocity caused by fittings like diverging sections, contracting sections, openings, and dampers.

Normally, dynamic pressure losses (Δp_d) are proportional to the velocity pressure and can be expressed as

$$\Delta P_d = C_o \cdot P_v \tag{6.2}$$

where C_0 = dynamic loss coefficient dependent on the geometry of the particular fitting,

P_v = velocity pressure ($1/2\ \rho v^2$, ρ is the density of air).

The value of C_0 is measured experimentally and is tabulated in reference books such as the ASHRAE Handbook of Fundamentals.

6.3 System Curves and Fan Performance

The two main types of fans used for transporting air are centrifugal fans and axial-flow fans. Centrifugal fans are also divided into a number of types such as backward-curved, forward-curved, radial blade, and tubular centrifugal. Out of these, forward-curved and backward-curved fans are the most commonly used in air distribution systems (Fig. 6.2).

Forward-curved centrifugal fans generally operate at relatively low speeds and are used for transporting high volumes at low static pressure. These fans are lower in cost and are preferred to backward-curved fans in applications where high static pressures are not encountered. However, the main disadvantage of these fans is that they have an "overloading" characteristic that results in the power consumption curve having a positive slope until 100 percent airflow, as compared to backward-curved fans, which have maximum power consumption at about 80 percent of flow. Since backward-curved fans are more costly, they are generally used for high static pressure applications while axial flow fans are used to transport high volumes of air at low or no static pressure drops.

Fans and fan systems are normally rated based on pressure and flow rate. The two parameters are dependent on each other as the airflow rate produced by a fan depends on the system pressure (the pressure against which it needs to work).

The system resistance of a ducting system is the total sum of all pressure losses encountered in the system due to coils, filters, ducting, dampers, and diffusers. The system curve is a plot of the system resistance encountered at different volumes of airflow, as shown in Fig. 6.3. The system resistance varies to the square of the airflow, and the curve is parabolic in shape.

Forward curved fan   Backward curved fan

Figure 6.2 Backward- and forward-curved centrifugal fans.

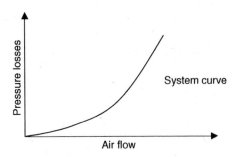

Figure 6.3 Typical system curve for an air distribution system.

Similarly, the relationship between the flow and pressure developed by a fan is called a fan curve. A fan curve shows all the different operating points of a fan at a particular operating speed as its discharge is throttled from zero to full flow. Since fans can operate at different speeds and with different impeller sizes, usually fan curves for different fan speeds and impeller sizes are plotted on the same axis. Figure 6.4 shows a typical fan curve for a fan operating at a particular speed and impeller size.

The actual shape of the fan curve depends on the type of fan used. The performance of some fans, such as axial-flow fans, also depends on the blade angle, which can be adjustable. For such fans, manufacturers provide a "family of curves" showing fan performance at different blade settings. Fan curves also include data such as the fan power required and operating efficiency at the different operating conditions, as shown in Fig. 6.5.

The system operating point is the point on the system resistance curve that corresponds to the required airflow condition. When a fan is selected for a particular application, a fan that has a performance curve that can intersect the system curve at the desired operating point is selected, as shown in Fig. 6.6.

As in the case of pumps, power consumed by fans is related to the product of the volume flow and pressure developed.

$$\text{Fan impeller power} \propto \frac{\text{air flow rate} \times \text{pressure developed}}{\text{efficiency}} \qquad (6.3)$$

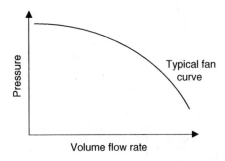

Figure 6.4 Typical fan curve.

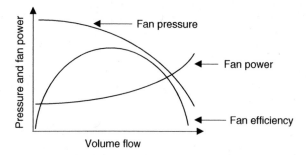

Figure 6.5   Typical fan curves provided by manufacturers.

In SI units:

Fan impeller power (kW)

$$= \frac{\text{flow rate (m}^3\text{/s)} \times \text{pressure developed (N/m}^2\text{or Pa)}}{1000 \times \text{efficiency}} \quad (6.4)$$

In Imperial units:

Fan brake horsepower

$$= \frac{\text{flow rate (cfm)} \times \text{pressure difference (in.water)}}{6350 \times \text{efficiency}} \quad (6.5)$$

Therefore, fan power consumption can be lowered by reducing the airflow rate, the pressure losses, or both, and will form the basis of some of the energy management measures for fan systems that are discussed later.

6.4 Affinity Laws

The performance of fans under different conditions are related by the fan affinity laws given in Table 6.1. The fan laws relate fan speed and impeller diameter to airflow, pressure developed by the fan, and impeller power.

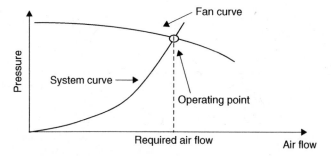

Figure 6.6   Matching of system and fan curves.

TABLE 6.1   Fan Affinity Laws

	Change in speed (N)	Change in impeller diameter (D)
Flow (Q)	$Q_2 = Q_1 \left(\dfrac{N_2}{N_1}\right)$	$Q_2 = Q_1 \left(\dfrac{D_2}{D_1}\right)^3$
Pressure (Δp)	$\Delta p_2 = \Delta p_1 \left(\dfrac{N_2}{N_1}\right)^2$	$\Delta p_2 = \Delta p_1 \left(\dfrac{D_2}{D_1}\right)^2$
Power (P)	$P_2 = P_1 \left(\dfrac{N_2}{N_1}\right)^3$	$P_2 = P_1 \left(\dfrac{D_2}{D_1}\right)^5$

Fan laws are used by manufacturers to estimate the performance of fans at different impeller sizes and operating speeds. Since the speed of fans can be easily changed by changing pulley sizes or by installing variable speed drives, the fan laws are very useful to system designers. They can be used to estimate the performance of fans at different speeds, as illustrated in Example 6.1.

Example 6.1   A fan delivers 5 m^3/s at 1400 rpm and consumes 20 kW. Calculate the new airflow rate and power consumption if the fan speed is reduced to 1200 rpm.

$Q_1 = 5 \ m^3$/s

$P_1 = 20$ kW

$N_1 = 1400$ rpm

$N_2 = 1200$ rpm

$Q_2 = Q_1 \times (N_2/N_1) = 5 \times (1200/1400) = 4.3 \ m^3$/s

$P_2 = P_1 \times (N_2/N_1)^3 = 20 \times (1200/1400)^3 = 12.6$ kW

This example also shows that when the speed is reduced by 14 percent (1400 to 1200 rpm), theoretically, the power consumption reduces by about 37 percent (20 to 12.6 kW). It should be noted that this calculation does not take into account the change in fan efficiency due to change in operating speed.

6.5   Energy Saving Measures for Air Handling and Distribution Systems

6.5.1   Ducting system design

The power required by a fan depends on the airflow it needs to deliver and the system pressure against which it has to work. While the airflow required in a cooling or heating system depends on the load that needs to be satisfied, the pressure losses are not directly related to the load. Hence, although the amount of air to be delivered by a system may not be totally within the control of a designer, the pressure losses can be minimized through ducting system design. As can be seen from Eq. (6.5), a reduction in pressure losses helps to directly reduce fan power. For example, if the pressure losses can be halved, the fan power can also be reduced by half.

Another point to remember is that a fan also adds heat to the air that it is moving. The temperature rise across the fan depends on the pressure rise across the fan and the fan efficiency. For a fan working at 70 percent efficiency, which is developing a total pressure of 500 Pa, the temperature rise can be about 0.5°C. In a typical air-conditioning system, the difference in temperature of the air being delivered to the space being cooled and the air leaving the space is about 10°C. For such a case, the heat added by the fan can be about 5 percent of the total cooling load. This percentage can be even higher at part load, when the supply air temperature is higher.

Therefore, in air-conditioned systems, it is very important to minimize the system pressure losses to enable fan power to be reduced while also helping to decrease the cooling load on the chillers.

Pressure losses occur in air distribution systems due to frictional losses and dynamic losses. As explained earlier, dynamic losses in ducting systems occur due to change in flow direction, caused by fittings like elbows, bends, and tees, and changes in area or velocity, caused by fittings like diverging sections, contracting sections, openings, and dampers. Usually, dynamic losses tend to be higher than frictional losses in ducting systems.

From Eq. (6.1), it is evident that the fan energy required to overcome frictional losses can be reduced by lowering the friction factor, duct length, and air velocity in ducts. The most commonly used materials for fabricating ducting are galvanized sheet metal and aluminium sheets, both of which are smooth and have low friction factors. However, in ducting systems, it is sometimes necessary to use flexible ducting or ducting with internal lining, which results in higher frictional losses. In such cases, the length of such ducts should be minimized as far as possible.

Frictional losses in ducts also depend on the velocity of airflow in the ducting system. From Eq. (6.1) it can be seen that higher velocity or a lower cross-sectional area (hydraulic mean diameter) result in higher losses. Therefore, to minimize losses, the duct cross-sectional area needs to be increased. However, selecting a duct with a large cross-sectional area to minimize frictional losses would lead to higher cost for ducting since ducting material and labor costs depend on size. As such, duct designing is matter of economics, which requires balancing the first cost and the running cost of the system. Based on experience, duct loss of 1 Pa/m is found to be a reasonable value to be assumed while designing ducting systems.

Similarly, dynamic losses can be minimized by using duct fittings that have lower loss coefficients. Sudden changes in direction should be avoided and where required, bends should be used rather than sharp elbows. Further, obstructions in ducts should be avoided and diverging and converging sections should be made gradual with the angle of divergence and convergence not exceeding 12° and 30°, respectively.

6.5.2 Fan discharge and inlet system effects

Fan inlet and outlet conditions also affect system losses, which can result in higher fan power to satisfy the requirements.

Fans impart dynamic pressure on the air due to centrifugal action. This dynamic pressure has to be converted to static pressure to enable the fan to overcome the system losses. Usually, a minimum duct length is required after the fan to enable this static regain to be completed. Therefore, a system design should attempt to use straight ductwork for three to five equivalent duct diameters downstream of the fan discharge. Where transition to a duct with larger area is required after the fan, a taper having an included angle of no more than 15° should be used, as shown in Fig. 6.7.

Where fans discharge into a plenum, losses occur due to the sudden enlargement in flow area. The addition of a short discharge duct of only one or two equivalent diameters in length significantly reduces this sudden enlargement loss.

Duct bends immediately at the fan discharge also create a high static pressure drop due to turbulence and the velocity profile existing at the discharge. If an elbow must be used at the fan discharge, it is recommended not to have one with a short radius. Preferably, an elbow with a minimum radius of 1.5 times the equivalent duct diameter should be used.

Similarly, fan inlet conditions also affect fan performance. Nonuniform flow into the suction of a fan is typically caused by an elbow installed too close to the fan inlet. This will not allow the air to enter the impeller uniformly and will result in turbulent and uneven flow distribution at the fan impeller. This results in lower fan efficiency and higher fan power. If it is necessary to have a sharp bend at the fan inlet, vanes and a splitter can be installed to minimize losses.

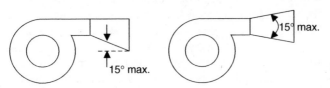

Figure 6.7   Fan discharge transitions.

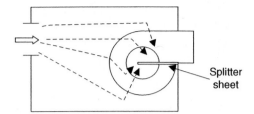

Figure 6.8 Splitter sheet for unbalanced flow.

In many air handling units, the opening is not vertically centered to the fan inlet. By adding a simple splitter sheet on each side of the fan, as shown in Fig. 6.8, the performance of the fan can be improved.

6.5.3 Filter losses

Various types of air filters are used in air handling units to filter the air before being supplied to the occupied spaces. These air filters remove suspended solid or liquid materials from the fresh air (required for ventilation) and the recirculation air.

The most common type of filters used for air handling equipment are media filters. These filters, which are normally made of fibrous material, trap particles while the air passes through them. Media filters offer a resistance to the airflow depending on the type of filter and the amount of airflowing through it. When these filters accumulate dust, the pressure drop across them increases. Usually, filters are selected to work up to a design pressure drop provided for in the design, after which they need to be cleaned or replaced. As shown in Fig. 6.9, the pressure drop across the filter is lower than the value assumed in the design when it is clean. This results in higher airflow (Q_3) when the filter is clean. Gradually, when the pressure drop increases, the airflow drops to the value provided for in the design (Q_2). If the filters are not cleaned or replaced at this stage, the higher pressure drop results in lower airflow (Q_1).

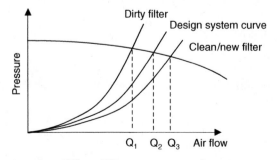

Figure 6.9 Effect of filter on system performance.

From an energy efficiency point of view, two aspects need to be considered in such filter operations. Firstly, when the filter is clean, more airflow than required is provided by the system. From Eq. (6.3) we know that the higher the airflow, the higher the energy consumption is. This means that, during the period from when the filter is clean to when it reaches the design pressure drop, if the airflow can be reduced, the energy consumed by the fan can be reduced. The easiest way to achieve this is by using a variable speed drive (VSD) to modulate the fan speed and provide the designed airflow, as shown in Fig. 6.10. Since the airflow delivered by the fan is proportional to the fan speed, the fan can be operated at speeds lower than the designed speed and gradually increased to the designed value. Since the fan power is proportional to the cube of the fan speed, significant energy savings can be achieved by this measure.

Secondly, energy savings can be achieved by replacing or cleaning the filter when the pressure drop reaches the maximum value allowed for in the design. A sensor can be installed to provide an alarm when the filter needs cleaning or replacing. If nothing is done to the filters at this stage, the airflow will drop and will result in insufficient airflow to some areas, causing discomfort to occupants in these areas. In extreme cases the problem of insufficient airflow is solved by running the fan at a higher speed by changing the pulleys or setting a higher set point on the VSD, if such a device is used. Both these solutions lead to higher fan power consumption.

When filters are not cleaned or replaced regularly, apart from resulting in higher pressure drops, dirt particles sometimes pass through them and lodge in the cooling coils. This results in a higher pressure drop across the coils, which leads to higher fan power (discussed later) and the need for more frequent cleaning of the coils, which is even more expensive than cleaning or replacing filters.

Another type of filter being used in air handling units are electronic air filters, which use "electrostatic precipitation" to effectively remove

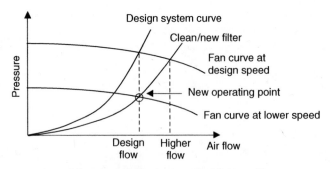

Figure 6.10   Effect of reducing fan speed with clean filter.

particles as small as 0.01 microns. Electronic air cleaners circulate air contaminated with particles through a series of ionizing wires and plates that generate positive ions, and a section of collection plates that precipitate the ionized particles out of the air. Electrons, which are randomly present in the air, accelerate rapidly towards the positively charged ionizing wires. On the way to the wires, these accelerating electrons strike electrons out of other air molecules, making them positive ions. The positive ions become attached to the pollutant particles and are collected in the collector section, which has a series of thin metal plates that are alternatively charged positively and negatively with a high DC voltage.

Unlike media filters, which have pressure drops of 25 to 50 Pa (sometimes even much higher), these electronic air cleaners offer little resistance to the airflow. Since, according to the fan affinity laws, fan pressure is proportional to the square of the fan speed and fan power is proportional to the cube of fan speed, fan power is proportional to the fan pressure raised to the power of 1.5 (i.e. kW $\propto$ pressure $^{3/2}$). Therefore, significant savings in fan power can be achieved by using electronic air cleaners.

Example 6.2 A system with a media filter has a fan that delivers 10 m^3/s at 500 Pa and consumes 10 kW. If the media filter has a pressure drop of 50 Pa, calculate the savings in fan power consumption that can be achieved if the media filter is replaced with an electronic air cleaner that has a negligible pressure drop.

$Q_1 = 10$ m^3/s

$P_1 = 10$ kW

$\Delta p_1 = 500$ Pa

$\Delta p_2 = (500 - 50) = 450$ Pa

$P_2 = P_1 \times (\Delta p_2/\Delta p_1)^{3/2} = 10 \times (450/500)^{3/2} = 8.5$ kW

Therefore, savings in fan power consumption will be 1.5 kW.

The fan power savings achieved depends on the actual pressure drop across the media filter, the total pressure drop in the system, and fan characteristics. However, it should be noted that to achieve this saving in a retrofit situation, it is necessary to reduce the fan speed to deliver the designed airflow rate after the media filter is replaced with an electronic air cleaner. This can be achieved through a pulley change on the fan drive or by using a variable speed drive to operate the fan at a lower speed. Otherwise, if nothing is done to reduce the fan speed after the electronic air cleaner is installed, the lower pressure drop will just result in higher airflow without any significant savings in fan power. This is a very important point to remember as sometimes vendors of electronic air cleaners project the feasibility of installing their devices based on fan power savings and the cost of the air cleaner, omitting the cost incurred by reducing fan speed.

Due to the high first cost of electronic air cleaners as compared to media filters, in many situations, it is not possible to justify their installation based on energy savings alone. In such cases, other benefits such as the reduction in periodic maintenance cost, associated with cleaning and replacing of media filters, and improved air quality should be included in the cost benefit analysis.

6.5.4 Coil losses

Clean coils.   Like filters, coils too offer significant resistance in air distribution systems. The pressure drop across coils depends on their design (density of rows and fins), face velocity, and how well the coils have been maintained (how clean).

Coils are placed downstream of filters so that any particles in the air can be filtered before the air reaches the coil. However, if the filters used are not efficient or well maintained they cannot remove dirt particles. This results in the coil acting as the filter. Since cooling coils are wet and consist of narrow passages for the air to pass through, anything that gets past the filters normally tend to get stuck on cooling coils. Cooling coils that have been in use for about 10 years and have not been cleaned regularly result a significant drop in airflow due to blockage. One way of checking the condition of the coil is by comparing the actual coil pressure drop with the designed coil pressured drop. However, if the design data is not available and the pressure drop across the cooling coil is over 250 Pa, chances are that it is partially blocked and needs cleaning.

As with filters, blocked coils result in lower airflow and can lead to discomfort for the occupants of the building if there is a substantial drop in coil performance. Often, when such problems are encountered, the fans are operated at higher speed to increase the airflow. This leads to higher fan power consumption.

Face velocity.   Apart from keeping the coils clean, fan power consumption can also be minimized through coil design and selection. The pressure drop across coils depends on the number of coil rows and fin density as well as the velocity of airflowing through them.

The face velocity of a coil is the velocity of air passing through the coil. The pressure drop across a coil is proportional to the square of the velocity. Therefore, if the face velocity can be reduced by 10 percent, the pressure drop can be reduced by about 20 percent. Reduction in coil face velocity can be achieved by making the coil bigger. However, this leads to higher first cost for the coil due to its larger size and also higher cost for the entire air handling unit since it has to be made bigger to accommodate the larger coil. Usually, to minimize first cost, coils are designed for face velocities of about 2.5 m/s. Although low face velocity coils with face velocities of about 1 m/s cost more at first, they have a much lower operating cost due to lower fan energy consumption. Therefore, if coils and air handling units are selected based on life-cycle costing, low face velocity coils offer significant energy savings.

Example 6.3   For a system delivering 10 m^3/s, the pressure drop across a cooling coil is 200 Pa and the face velocity is 2.5 m/s. Calculate the pressure drop and the reduction in fan power if the face velocity is reduced to 1.0 m/s for this application. Assume the fan efficiency is 70 percent.

$Q = 10$ m^3/s

$v_1 = 2.5$ m/s

$v_2 = 1.0$ m/s

$\Delta p_{1\ (coil)} = 200$ Pa

$\Delta p_{2\ (coil)} = \Delta p_{1\ (coil)} \times (v_2/v_1)^2 = 200 \times (1.0/2.5)^2 = 32$ Pa

Reduction in pressure drop across the coil = $(200 - 32) = 168$ Pa
From Eq. (6.4), reduction in fan power = $[10\ (m^3/s) \times 168\ (Pa)]/[1000 \times 0.7] = 2.4$ kW.

Dual-path air handling units.   Deep coils with high fin density are normally required for dehumidification applications. Such coils not only have high pressure drops but are also difficult to maintain. In conventional air handling units (Fig. 6.11), outdoor air provided for ventilation is mixed with the return air before passing through the cooling coil. The outdoor air, with higher moisture content is mixed with a large quantity of return air with lower moisture content. Since the quantity of outdoor air is much less than the quantity of return air, this results in a mixture of air with a lower overall moisture content. This makes dehumidification harder as the potential for moisture removal is reduced and leads to the need for deep coils.

This shortcoming can be overcome by using dual-path air handling units (Fig. 6.12). In such air handling units, two separate coils are used to treat the outdoor air and return air before mixing them. The coil for

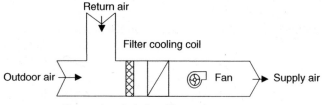

Figure 6.11   Conventional air handling unit.

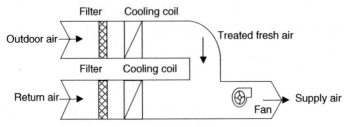

Figure 6.12 Dual-path air handling unit.

treating the outdoor air is designed for carrying out most of the dehumidification while the return air coil is sized for mainly providing sensible cooling. The two air streams are then mixed to give the desired supply air conditions.

Since the quantity of outdoor air for dehumidification is small and the moisture content is relatively high, low face velocity coils can be used for treating outdoor air. This leads to lower coil pressure drop and reduced energy cost. Similarly, since the return air coil mainly performs sensible cooling, a shallow coil can be selected for this application. This too results in lower coil pressure drop and helps reduce fan power consumption.

6.5.5 Fan efficiency

Normally, centrifugal fans used in air distribution systems are forward-curved or backward-curved (Fig. 6.2).

Forward-curved fans rotate at relatively slow speeds and are generally used for producing high volumes at low static pressure. The maximum static efficiency of forward-curved fans is usually between 60 to 70 percent and occurs at less than 50 percent of the maximum airflow, as shown in Fig. 6.13. The fan power curve has an increasing slope and is referred to as an "overloading type." The advantage of these fans is that they are relatively low cost.

Backward-inclined fans operate at about twice the speed of forward-curved fans. The maximum static efficiency is higher and is about 80 percent, and occurs at about 60 to 70 percent of the maximum flow, as shown in Fig. 6.13. The advantage of backward-curved fans is their higher efficiency and the nonoverloading characteristic of their power curve. However, they are normally more costly than forward-curved fans.

Due to the lower first cost, forward-curved fans are commonly used in air handling units (AHUs). However, from Eq. (6.3) it can be seen that fan power can be reduced by increasing fan efficiency. In a system where the airflow and pressure is fixed, a 10 percent improvement in fan efficiency will lead to an equivalent reduction in fan power consumption. Therefore, although a backward-curved fan may have a higher first cost,

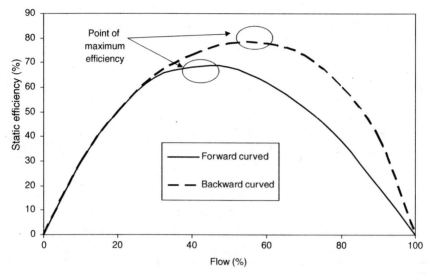

Figure 6.13 Fan efficiency for forward-curved and backward-curved fans.

based on a life-cycle cost, it may be much more attractive as the extra cost of the fan may be recovered by the lower fan power consumption during operation.

However, it should be noted that fan efficiency also depends on the operating point, as shown in Fig. 6.14. Although a backward-curved fan may have a maximum static efficiency of 80 percent, it may operate at 60 percent static efficiency at the operating point. Therefore, when a

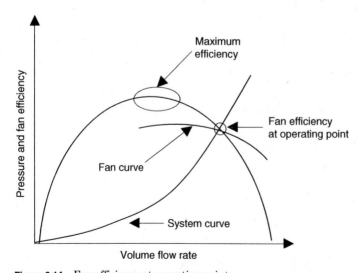

Figure 6.14 Fan efficiency at operating point.

fan is selected for an application, it should be selected so that it operates at its highest efficiency at the desired operating point.

6.5.6 Excess airflow

In air handling systems and ventilation systems, fans are selected to provide the airflow rate required for maintaining space conditions. This
designed airflow rate is normally computed based on expected cool-
ing/heating and ventilation loads. Since the airflow delivered by a fan
also depends on the pressure losses in the system, which it has to overcome to deliver the required airflow rate, the system pressure losses too
need to be estimated prior to selecting a fan.

As described in Section 6.2, system losses depend on factors such as
friction losses in ducting, losses due to fittings, and changes in velocity
and direction. In system design, usually, a safety factor is added to account
for differences between computed values and actual system losses that
may result due to reasons such as changes in ducting layout during
installation to overcome site constraints. The safety factor used for designing an air distribution system usually depends on how confident the
designer is about the design. It is not uncommon to see systems designed
with high safety factors, which result in excess airflow during actual
operation due to the intersection of the fan curve and actual system curve
at a point different from the design operating point (Fig. 6.15).

As explained earlier in Eq. (6.3), fan power depends on, both, the airflow delivered and the system pressure losses. Therefore, excess airflow
results in higher fan power consumption.

The best way to check whether the airflow is excessive is to measure
the actual airflow and compare with the design requirements. If design
data is unavailable, then a good indication of excessive airflow can be provided by dividing the AHU airflow by the floor area it serves. Although
the airflow provided per unit floor area varies with design conditions

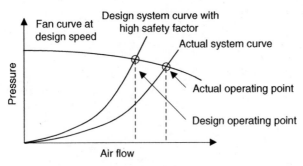

Figure 6.15 Fan performance due to use of high safety factor
for estimating system losses.

such as ceiling height and supply air temperature, generally, for cooling the ratio is in the range of 18 to 27 cmh/m^2 (1 to 1.5 cfm/ft^2). Therefore, if the actual ratio of airflow to floor area served by the AHU is higher than this, it is very likely that the operating fan capacity is more than required.

If the airflow is found to be excessive, the simplest solution is to reduce the fan speed since airflow is proportional to the fan speed (Table 6.1). This can be easily achieved by changing the pulley sizes of the fan and the motor.

Since reduction in fan speed is proportional to the required reduction in airflow, if the airflow needs to be reduced by 20 percent, then the fan speed also needs to be reduced by 20 percent. The pulley sizing to achieve this reduction in fan speed is illustrated in Example 6.4.

Example 6.4 A system is found to be delivering 12 m^3/s when the actual requirement is 10 m^3/s. The fan speed is measured to be 800 rpm while the motor speed is 960 rpm. The motor pulley size is 200 mm and the fan pulley size is 240 mm. Find the new fan pulley size and motor power consumption.

$Q_1 = 12$ m^3/s

$Q_2 = 10$ m^3/s

$N_1 = 800$ rpm

New fan speed, $N_2 = N_1 \times (Q_2/Q_1) = 800 \times (10/12) = 666$ rpm

Since the fan and motor speeds and pulley diameters are related as follows:

$$N_{(fan)}/N_{(motor)} = D_{(motor)}/D_{(fan)}$$

New fan pulley diameter, $D_{(fan)} = (D_{(motor)} \times (N_{(motor)})/N_{(fan)}$

Therefore, $D_{(fan)} = 200 \times (960/666) = 288$ mm

Another common way of reducing fan speed is by using a VSD for the fan motor and then running the motor at a lower speed. Although this solution is normally more expensive than a pulley change, it is easier to achieve the desired airflow through trial-and-error setting of the VSD's frequency to get the exact airflow requirements.

6.5.7 Type of air distribution system (CAV versus VAV)

Fans for air handling units are normally sized to handle the maximum airflow required to meet peak load conditions. However, peak load conditions are usually experienced only for short periods of time and the capacity of air handling units is controlled to match requirements by varying the supply air temperature or the amount of air supplied.

In constant air volume (CAV) systems, the capacity is controlled by varying the supply air temperature. In such systems, the fan is operated at a fixed speed to give a fixed quantity of air. This not only wastes energy by supplying a constant volume of air irrespective of the load, but also leads to high space relative humidity in air-conditioning systems at low loads due to higher operating supply air temperatures at part load.

To avoid these shortcomings, variable air volume systems with devices such as discharge dampers, inlet guide vanes, or variable speed drives can be used to regulate the air volume with load while maintaining a fixed supply air temperature. Although discharge dampers and inlet guide vanes are able to reduce the air volume, the energy savings achieved are much less than for variable speed drives, which are able to closely follow the theoretical "cubic" fan power relationship. Typical energy savings achieved by varying the airflow in VAV systems using the different systems is illustrated in Fig. 6.16.

A VSD can vary the speed of the fan to produce different airflows by changing the fan curve. As shown in Fig. 6.17, at part load conditions, the fan speed can be reduced so that the resulting fan curve intersects the system curve at the required operating point. Variable speed fans are able to closely match the theoretical fan power consumption curve, which is dependent on the cubic relationship $kW \propto (flow)^3$. Therefore, theoretically, at 50 percent flow condition, the fan power consumption is 12.5 percent $(0.5^3 = 0.125)$ of the maximum power consumption. This shows that it is possible to reduce the power consumption significantly at part load conditions by using a variable speed drive.

6.5.8 Static pressure set point

In variable air volume (VAV) air distribution systems, the air supplied to the different spaces is varied based on load requirements. This is

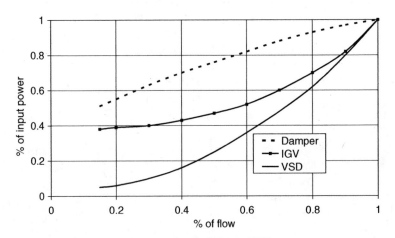

Figure 6.16 Fan energy consumption in different VAV systems.

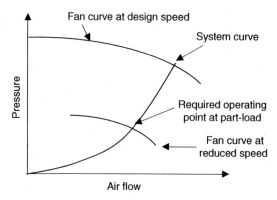

Figure 6.17 Reducing air flow at part load by reducing fan speed.

achieved by using VAV boxes that have dampers to automatically control the airflow through each box. Each VAV box receives an input from a thermostat, which monitors the respective space temperature. The measured space temperature is continuously compared with the desired space temperature set point and a signal is provided to automatically vary the airflow through each VAV box by opening or closing its damper.

Therefore, depending on the load, the VAV boxes vary the quantity of air supplied to the spaces being conditioned. When the load is high, more air is provided, while at part load less air is provided. This causes the static pressure in the distribution ducting system to vary with load. In VAV systems, static pressure sensors are installed in the distribution ducting to sense the load and the pressure signal used to control the airflow from individual AHU's by varying the fan speed or adjusting the inlet guide vanes to maintain a static pressure set point, as shown in Fig. 6.18.

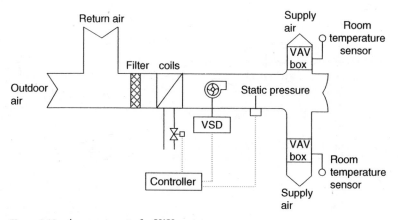

Figure 6.18 Arrangement of a VAV system.

Since the static pressure in ducts is used to control the airflow based on load requirements, the location of the static pressure sensors and actual set point are important to maximize energy savings. If the duct static pressure sensor is not located correctly or the static pressure set point is too high, the system will operate at a much higher pressure than required and will result in wasted energy. It is very common to find the duct static pressure sensor located at the discharge of AHUs. In such systems, since the duct pressure sensor is unable to sense the lowest pressure in the system, the pressure set point has to be set higher to ensure that sufficient pressure is maintained in all areas of the system.

The static pressure sensor should be located in the ducting system in the region where the lowest static pressure is expected. The location of this point depends on the design of the ducting system, but is usually about two-thirds of the way along the duct. The actual location can be found by measuring the duct pressure at a few locations along the ducting system.

The static pressure set point is important because it ensures that the system performs by providing sufficient air to all areas (depending on load) while saving energy during part load operation. A high-pressure set point will ensure that the system airflow requirements of all areas are maintained but will result in lower energy savings. Similarly, a low-pressure set point may result in higher energy savings but may lead to space comfort conditions not being met in some areas. Therefore, it is essential to set the most optimum set point to maximize energy savings while maintaining space comfort conditions in all areas served by the system. This ideal set point can normally be found by trial-and-error, by operating the system at different set points and monitoring the space conditions.

6.5.9   VAV optimization algorithm

In VAV systems, further energy savings can be achieved by continuously resetting the static pressure set point based on system operations. In such systems, the static pressure set point is continuously varied (as opposed to a fixed set point in conventional systems) in response to the damper positions of the VAV boxes to ensure that no box is starved of air. The aim of such a system is to minimize AHU fan energy consumption by reducing the static pressure in the system while ensuring that the space cooling or heating requirements of all spaces are met. However, for such a system to function, the VAV box damper positions should be monitored and be linked to the AHU fan controls through a central BAS or EMS.

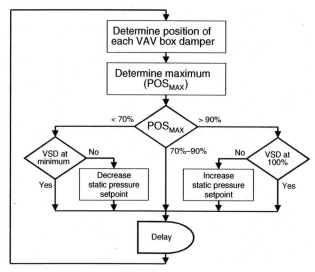

Figure 6.19 Typical static pressure reset algorithm.

A typical static pressure reset algorithm is shown in Fig. 6.19. The control algorithm is set to reduce the AHU fan speed up to the minimum set point while ensuring no VAV box is more than 90 percent open or likely to be starved of air. The algorithm is able to achieve this by monitoring the damper positions of all the VAV boxes in a particular system and identifying the box with the damper open the most and, thereby, the most likely box to be starved of air. If the box with the damper open most is more than 90 percent open, then the static pressure set point is raised. Similarly, if the box with the damper open most is open less than 70 percent (other boxes are open even less), the static pressure set point is reduced. This check is performed continuously by the control system to ensure that the system pressure is maintained at the minimum possible. It is estimated that such systems are able to achieve further 20 percent energy savings as compared to conventional fixed set point VAV systems.

6.5.10 Air distribution and balancing

In central systems, air is treated normally in AHUs and distributed to one or more of the conditioned spaces. The cooling or heating provided to meet space requirements depends on the quantity of air supplied and the temperature of the supply air. As the temperature of the air supplied by a particular AHU to the different spaces it is serving is the same, different quantities of air have to be supplied to each of the spaces as they may have varying load requirements. Since treated air from an AHU is distributed by a ducting system to the different spaces, the ducting system needs to be able to distribute the airflow so that each space receives the required quantity of air. This is normally achieved by air balancing.

If airflows to the different areas are not balanced, some areas may receive more airflow while others receive less airflow. This normally results in hot and cold spots in air-conditioned spaces. Often, "hot spots" are experienced in areas furthest from AHUs due to insufficient airflow at such locations. If simultaneously, "cold spots" are experienced in areas closer to AHUs, this would indicate that the airflow is not well balanced.

In such situations, rather than increasing the airflow supplied by the AHU fan to satisfy warm areas (which results in higher energy consumption), the airflows should be balanced so that excess air can be diverted from areas receiving too much flow to areas that have inadequate flow.

6.5.11 Runaround coils

For areas that require low space relative humidity, usually, the supply air is "overcooled" by the AHU coil to remove sufficient moisture and thereafter electric duct heaters are used to reheat the supply air before releasing it into the occupied space. This strategy not only wastes energy but also increases the cooling load on the chiller plant.

Reheating for humidity control can be avoided by using runaround coils where two extra coils sandwich the main cooling coil of the AHU (Fig. 6.20). Water is circulated through the two coils using a small pump, which enables transferring of heat from the incoming air (pre cooling) to the leaving air. Heat pipe systems can also be used to achieve the same effect. Research has shown that runaround coils can achieve substantial

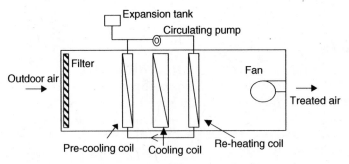

Figure 6.20 Arrangement of an AHU runaround coil.

energy savings (about 20 percent of the total annual cooling energy for the cooling coil) in hot and humid climates.

6.5.12   Optimal start-stop

The optimal start-stop algorithm available on most building automation systems (BAS) can predict how long a building or space will take to reach the desired temperature based on the variables that affect it, such as outdoor air temperature, indoor space temperature, and building thermal characteristics. This algorithm can be used to start the AHUs (and chillers/boilers) at the latest possible time to achieve the required space conditions before the space is occupied. Similarly, at the end of the day, the algorithm can help to shut down the cooling or heating plant at the earliest possible time.

6.5.13   Space temperature reset

Air-conditioning systems maintain comfort conditions in occupied spaces by removing heat and moisture generated by occupants and equipment. Occupants in air-conditioned spaces reject heat to the surroundings by sensible and latent heat transfer to maintain body thermal balance. Sensible heat transfer is mainly by convection and radiation while latent heat transfer is by evaporation of moisture from the body. Therefore, the amount of heat rejected by an occupant into a space depends on parameters such as space temperature, relative humidity, and air circulation rate. An occupant is able to reject more latent heat if the space relative humidity is low. Similarly, more heat rejection takes place by sensible means if the space temperature is lower. For example, if the relative humidity is low, the occupants feel comfortable at a higher space temperature since less sensible heat transfer is required to maintain the body's thermal balance. The required space comfort conditions also depend on factors such as occupant activity and their clothing. Therefore, the space temperature should be set taking into account factors such as space relative humidity, occupant activity, and clothing level.

The space temperature set point for comfort applications also depends on the heat gain into a space by radiation. For spaces with glazing, located in perimeter zones of buildings, the space temperature is normally set lower to account for the radiant heat transfer between the surroundings and the occupants. However, in such spaces at times when the mean radiant temperature is lower (evenings, night time or rainy days), occupants often feel cold if the normal space temperature set point is maintained. Therefore, to avoid such situations, the space temperature set point could be reset based on outdoor weather conditions.

6.5.14 Economizer cycle

An energy saving feature that can be incorporated into AHUs in some climates is the outdoor air economizer. The basis of this strategy is to use 100 percent outside air when it is below a certain temperature to cool the space rather than using a mixture of outside air and return air.

When the outdoor air dry-bulb temperature is below the indoor temperature, the economizer cycle can be programmed to convert the AHU to use 100 percent outdoor air by adjusting the position of the outdoor air and return air dampers. In humid climates, it is better to use enthalpy (sensible and latent energy level) based controls to activate this energy saving strategy.

A typical arrangement of an AHU working on an economizer cycle is shown in Fig. 6.21. An economizer system normally includes indoor and outdoor temperature sensors, motorized dampers, and controls. Since the outdoor air damper needs to be large enough to provide 100 percent outside air, this becomes a constraint when fitting economizers to existing AHUs.

Example 6.5 illustrates how the savings in cooling load, achievable through an economizer cycle, can be estimated.

Example 6.5 A building AHU working on the economizer mode provides 0.5 m³/s of outdoor air to replace return air at 22°C dry-bulb temperature and 65 percent relative humidity. If the conditions of the outdoor air are 18°C dry-bulb temperature and 45 percent relative humidity, estimate the reduction in cooling load for the building chillers. Take the density of air to be 1.2 kg/m³.

From the psychrometric data in Appendix B, the enthalpy of return air is 49.3 kJ/kg (at 22°C dry-bulb at 65 percent RH) and the enthalpy of outdoor air is 32.6 kJ/kg (at 18°C and 45 percent RH).

$$\text{Reduction in cooling load} = \text{airflow rate} \times \text{density of air} \times \text{difference in enthalpy}$$

$$= 0.5 \text{ m}^3/\text{s} \times 1.2 \text{ kg/m}^3 \times (49.3 - 32.6) \text{ kJ/kg}$$

$$= 10 \text{ kW (approximately 2.8 RT)}$$

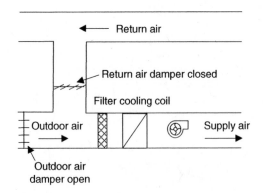

Figure 6.21 Typical arrangement of an AHU on economizer mode.

6.5.15   Fresh air control

Outdoor air is required for ventilation and to ensure sufficient building pressurization to prevent infiltration of air into buildings. To maintain good building Indoor Air Quality (IAQ), ASHRAE Standard 62 is generally used as a guideline for determining the quantity of ventilation air required per occupant.

The quantity of fresh air required to provide sufficient ventilation depends on factors such as the number of occupants and the type of space usage. Usually, the amount of fresh air provided works out to be about 10 to 15 percent of the total supply air quantity. Therefore, if the fresh air intake is more than about 15 percent of the total supply air quantity, it could indicate an excess supply of fresh air. Reduction in the amount of fresh air provided would help to reduce the cooling and heating load of buildings and, therefore, result in lower energy consumption, as illustrated in Example 6.6.

Example 6.6   If the outdoor air is at 32°C (dry bulb) and 70 percent relative humidity, estimate the savings that can be achieved in cooling load when the total outdoor air intake for a building is reduced by 1 m³/s. Take the condition of room air to be 23°C (dry bulb) and 55 percent relative humidity.

From the psychrometric data in Appendix B, the enthalpy of outdoor air is 86.1 kJ/ kg (at 32°C dry bulb and 70 percent RH), and the enthalpy of return air is 47.5 kJ/kg (at 23°C and 55 percent RH).

$$\text{Reduction in cooling load} = \text{airflow rate} \times \text{density of air}$$
$$\times \text{difference in enthalpy}$$
$$= 1.0 \text{ m}^3/\text{s} \times 1.2 \text{ kg/m}^3 \times (86.1 - 47.5) \text{ kJ/kg}$$
$$= 46.3 \text{ kW (approximately 13 RT)}$$

Further, as building occupancy also usually varies during different periods of the day, it is not necessary to provide a constant amount of fresh air at all times. Hence, fresh air provided can be varied based on building occupancy.

One way to achieve this is to preprogram the expected occupancy pattern on the building energy management system and use it to vary the outdoor airflow. A simpler solution is to use carbon dioxide (CO_2) sensors to monitor the CO_2 level in the occupied spaces or the return air at the AHUs and use it as an indicator of occupancy to vary the outdoor airflow. The CO_2 level can be set at a predetermined value, such as 1000 ppm, and the proportion of the outdoor and return airflow quantities varied to maintain this set point, as shown in Fig. 6.22.

6.5.16   Air-to-air heat recovery

As described earlier, ventilation in buildings is provided by introducing outdoor air and exhausting an equal quantity of air to maintain the pressure balance in the building.

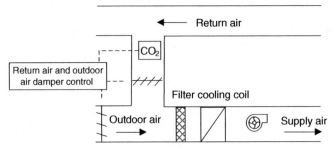

Figure 6.22   Outdoor air control based on CO_2 level.

In air-conditioned buildings, the exhaust air is normally colder and has a lower enthalpy than the fresh outdoor air that is taken in for ventilation. Therefore, energy recovery systems, such as those using heat pipe systems and energy recovery wheels (Fig. 6.23), can be used to precool the fresh air using exhaust air. In centralized ventilation systems, outdoor air is taken at one point and distributed to the individual AHUs. Similarly, the space air removed from the different parts of a building is also discharged using a single exhaust fan. Very often, both the fresh air and exhaust air fans are located on the roof of the building, as in the case of office towers. The exhaust air can then be easily used to precool the outdoor air, as shown in Fig. 6.24. The same can be done in heating systems to preheat fresh air using the warm exhaust air.

The amount of energy recovered depends on the efficiency of the recovery system and can be expressed as follows for systems that exchange sensible heat and those that can exchange total heat (both sensible and latent).

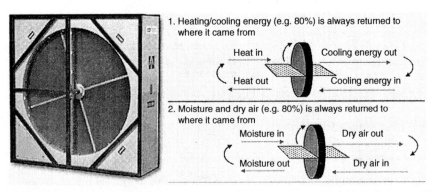

Figure 6.23   Arrangement and operation of a typical energy recovery wheel. *(Courtesy of Desiccant Rotors International Pvt. Ltd.)*

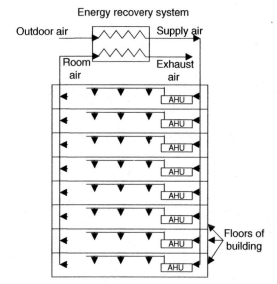

Figure 6.24 Typical arrangement of an energy recovery system.

Sensible heat only:

$$\eta_{sensible} = \left(\frac{T_{OA} - T_{SA}}{T_{OA} - T_{RA}}\right) \times 100 \tag{6.6}$$

Total heat:

$$\eta_{total} = \left(\frac{h_{OA} - h_{SA}}{h_{OA} - h_{RA}}\right) \times 100 \tag{6.7}$$

where η = efficiency (%)
   T = dry-bulb temperature (°C)
   H = enthalpy
   OA = outdoor air
   SA = supply air
   RA = return air

Similarly, the amount of energy recovered can be expressed as follows:

$$Q_{sensible} = \rho \times v \times Cp \times (T_{OA} - T_{SA}) \tag{6.8}$$

and

$$Q_{total} = \rho \times v \times (h_{OA} - h_{SA}) \tag{6.9}$$

where Q = energy recovered (kW)
　ρ = density of air (kg/m³)
　V = airflow rate (m³/s)
　Cp = specific heat capacity of air (kJ/kg·K)

Based on the values for specific heat capacity and density, Eq. (6.8) and Eq.(6.9) can be simplified as follows:

$$Q_{sensible} = 1.232 \times v \times (T_{OA} - T_{SA}) \qquad (6.10)$$

and

$$Q_{total} = 1.2 \times v \times (h_{OA} - h_{SA}) \qquad (6.11)$$

Example 6.7 illustrates how the amount of energy recovery can be computed.

Example 6.7   Consider the case where 1.2 m³/s of outdoor air is provided to an air-conditioning system to makeup for an equal amount of air removed by the exhaust system. If the outdoor air temperature is 30°C and the return air temperature is 23°C, find the supply air temperature that can be achieved using an energy recovery system that can only transfer sensible heat and has an efficiency of 75 percent. Also, calculate the total amount of precooling done to the outdoor air.
　Using Eq. (6.6),

$$T_{SA} = T_{QA} - (T_{QA} - T_{RA}) \times \eta_{sensible}$$

$$= 30 - (30 - 23) \times 0.75$$

$$= 24.75°C$$

Sensible cooling done:

$$Q_{sensible} = 1.232 \times v \times (T_{OA} - T_{SA})$$

$$= 1.232 \times 1.2 \times (30 - 24.75)$$

$$= 7.8 \text{ kW (which is approximately equal to 2 RT)}$$

6.5.17   Car park ventilation systems

Mechanical ventilation systems, consisting of supply and exhaust fans, are widely used for ventilating building spaces such as basement car parks. These ventilation systems are normally designed to provide the ventilation rates required under extreme or worst case conditions. However, ventilation rates required under normal operating conditions are usually much less and it is possible to control the operation of the fans to match the actual ventilation requirements.

For car parks, carbon monoxide (CO) and temperature sensors can be used to monitor the quality of car park air and control the supply and exhaust fans. The values for temperature and CO level can be set based on individual requirements or ventilation codes. Since the operation of exhaust and supply fans normally need to be interlocked to ensure a pressure balance in the areas served by them, the control system would need to control both sets of fans simultaneously.

In car park ventilation systems that have many supply and exhaust fans serving specific areas of the car park, sensors installed in various parts of the car park can be used to switch on/off the sets of supply and exhaust fans serving specific areas when the CO level and temperature in a particular part of the car park reaches a set value.

6.6 Summary

Air handling and air distribution systems are used for providing heating, cooling, and ventilation requirements in buildings. In such systems, fans are used to transport air through ducting systems from one part of a building to another. Since in a typical building, many fans are used for such applications, collectively they can account for a significant amount of energy consumed in buildings.

This chapter provided an overview of air handling and distribution systems, which included losses in ducting systems and fan characteristics. Thereafter, various design and operational strategies for reducing energy consumption by improving the energy efficiency of air handling and air distribution systems were described. The savings achievable for some of the energy saving measures were illustrated using examples.

Review Questions

6.1. A fan delivers 15 m^3/s at 1400 rpm and consumes 30 kW. If the airflow rate is to be reduced to 10 m^3/s, what will the new fan speed and power consumption be?

6.2. The fan of an air handling unit is designed to deliver 20 m^3/s. Estimate the savings that can be achieved in fan power consumption if the AHU uses a filter that has a pressure drop of 30 Pa instead of a filter having a pressure drop of 80 Pa.

6.3. If a building uses outdoor air at 35°C (dry-bulb) and 60 percent relative humidity, estimate the savings that can be achieved in cooling load when the total outdoor air intake for a building is reduced by 1.5 m^3/s. Take the condition of room air to be 23°C (dry-bulb) and 55 percent relative humidity.

If the building operates for 12 hours a day and 250 days a year, estimate the annual energy savings (in kWh) that can be achieved from the chiller plant if the efficiency of the chiller system is 0.8 kW/RT.

6.4. 3 m³/s of outdoor air is provided to an air-conditioning system to makeup for an equal amount of air removed by the exhaust system. If the outdoor air temperature is 33°C and the return air temperature is 23°C, find the supply air temperature that can be achieved using an energy recovery system that can only transfer sensible heat and has an efficiency of 77 percent.

Also, estimate the resulting annual energy savings from the chillers if the building operates 10 hours a day, 260 days a year, and the chiller system efficiency is 0.83 kW/RT.

Lighting Systems

7.1 Introduction

Lighting systems normally account for more than 20 percent of the electrical energy consumed in commercial buildings. Lighting systems not only consume power directly to generate light, in air-conditioned buildings they also indirectly account for some of the power consumed by air-conditioning systems, as the heat added by lighting has to be removed by the building cooling systems.

However, lighting is essential for buildings to ensure the comfort, productivity and safety of the building's occupants. Therefore, lighting systems need to be carefully designed to achieve the desired illumination level while using the minimum amount of energy.

Energy savings from lighting systems can be achieved by means such as optimizing lighting levels, improving the efficiency of lighting systems, using controls, and daylighting (using natural light). This chapter provides a brief description of some basic concepts of lighting followed by typical energy saving measures for lighting systems.

7.2 Definitions

A few important definitions associated with lighting systems are described here.

Lumens. Lumens is the SI unit for luminous flux, which is the quantity of light emitted by a source or the quantity of light received by a surface.

Typical values of luminous flux emitted by some common sources of light are given in Table 7.1.

TABLE 7.1   Luminous Flux Emitted by Common Light Sources

Lamp	Lamp wattage	Lumens
Torch lamp	3 W	30
Incandescent lamp	75 W	950
Compact fluorescent lamp	15 W	810
Fluorescent lamp	36 W	2,400
High-pressure sodium lamp	100 W	10,500
Low-pressure sodium lamp	131 W	26,000

Candela.   Candela (cd) is a measure of luminous intensity. Originally luminous intensity was measured in units called candles (based on the approximate amount of light emitted by a candle flame). Later the term candela was adopted to allow for consistent and repeatable measurements of light, where 1 candela is equal to 1 candlepower.

Lux.   Lux is the SI unit for illuminance, which is a measure of the direct illumination on a surface area of one square metre. One lux is one lumen/m^2. Some typical lux values are given in Table 7.2.

Luminous efficacy.   Luminous efficacy is the ratio of luminous flux emitted by a lamp to the power consumed by the lamp and its control gear. This ratio indicates the efficiency of a lamp in converting electrical power into light. The units of efficacy are lm/W.

Edison's first electric filament lamp had an efficacy of 1.4 lm/W. However, with research and development, the efficacy of lamps has improved significantly over the years. Typical values of efficacy for some common lamps are given in Table 7.3.

Color temperature.   The color temperature of a light source is a numerical measurement of its color appearance. It is based on the fact that when an object is heated to a temperature high enough it will emit light and as the temperature is increased, the color of the light emitted will also increase. For example, when a blacksmith heats a horseshoe, it will first appear red and will change to orange, followed by yellow and later white.

Color temperature is defined as the temperature of a blackbody radiator which emits radiation of the same chromaticity as the lamp. The

TABLE 7.2   Typical Lux Values

Location	Lux level
Basement car parks	15
Offices	500
Under the shade of a tree	10,000
Under the midday sun	100,000

TABLE 7.3   Typical Efficacy of Lamps

Lamp type	Efficacy (lm/W)
Incandescent	10–15
Halogen	13–25
Compact fluorescent	50–60
Fluorescent lube	69–100
Metal halide	85–120
High-pressure sodium	80–140
Low-pressure sodium	150–200

unit of color temperature is Kelvin (K). The degree of "warmth" or "coolness" of the space is related to the color temperature of the light source. The lower the color temperature, the "warmer" the light appears. Light sources that appear violet or blue color are "cool" while those that are red, yellow or orange are "warm."

Typical values of color temperature and associated warmness or coolness are given in Table 7.4.

Color rendering.   While color temperature is a measure of the color of a light source, the color rendering index is an indication of the ability of a light source to accurately show colors.

Color rendering expresses the appearance of object colors when illuminated by a given light source as compared to its appearance in a reference light source. It is usually expressed as an index called the color rendering index (CRI), which is an indication of the appearance of an object illuminated by a light source compared to its appearance under natural light. Natural light will have a CRI of 100. Electric filament lamps produce a continuous spectrum with all colors present and, therefore, they have a CRI of 100. Normally, CRI below 80 is considered poor color rendering while CRI above 80 is considered good.

Typical values of CRI are given in Table 7.5.

TABLE 7.4   Color Temperature and Warmness of Common Types of Lamps

Lamp type	Color temperature (K)
Incandescent filament lamp	2600–3000
Tungsten halogen	3000–3400
Warm white fluorescent	3000
Cool white fluorescent	4000
Daylight fluorescent	5000
Metal halide	3300–5700
High-pressure sodium	2000–3200
Low-pressure sodium	1600

TABLE 7.5   Typical Values of Color Rendering Index

Lamp type	Color rendering index (CRI)
Incandescent filament lamp	100
Tungsten halogen	100
Fluorescent	80–95
Metal halide	65–80
High-pressure sodium	25
Low-pressure sodium	0

7.3   Types of Lamps

7.3.1   Incandescent lamps

Incandescent lamps (Fig. 7.1) produce light using an incandescent fila-ment (normally tungsten), which is sealed in a glass bulb containing an inert gas.

Incandescent lamps were previously one of the most commonly used lamps due to their low initial cost and good color rendering (CRI above 95). However, they have been phased out from general applications due to their low efficacy, which is only about 10 to 15 lm/W. The low efficacy is because incandescent lamps convert about 92 percent of the energy into heat and only 8 percent is converted to useful light. They also have a relatively short life of only about 1000 hours.

However, they have good color rendering and a warm color tempera-ture, which make them the preferred choice for many special applications.

Figure 7.1 Incandescent lamp.
(*Courtesy of Philips.*)

Figure 7.2   Halogen lamp with reflector. (*Courtesy of Philips.*)

7.3.2 Halogen lamps

Halogen lamps are filament lamps (Fig. 7.2) that operate at higher pressure and temperature than standard incandescent lamps to produce whiter light. They have halogen added to prevent evaporated tungsten from blackening the bulb. This helps the lamps to maintain lumen levels and provide a "sparkling" effect.

They have an efficacy of 13 to 25 lm/W, which is slightly higher than that of incandescent lamps. They are commonly used in retail applications, such as for highlighting merchandise.

7.3.3 Fluorescent lamps

Fluorescent lamps are different from filament lamps and use an arc created between two electrodes in a tube filled with a gas or vapor (mercury) to produce light. The arc ionizes the gas and releases electromagnetic radiation in the ultraviolet region. The ultraviolet energy of the discharge activates the phosphor coating, which then emits light. They have higher efficacy than filament lamps. Fluorescent lamps can be

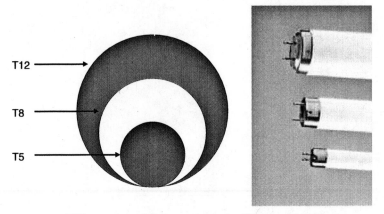

Figure 7.3   Different sizes of fluorescent lamps. (*Courtesy of Philips.*)

generally categorized into linear or tubular fluorescent lamps and compact fluorescent lamps.

Linear fluorescent lamps.   They are one of the most common types of lamps used for general lighting. They come in varying lengths. The tubes are classified by a T number (Fig. 7.3), which refers to the tube diameter in 1/8 of an inch. T12 refers to $12 \times 1/8 = 1$-inch (38 mm) tubes, while T8 refers to 1-inch or 26-mm tubes. The color rendering index for linear fluorescent lamps is normally about 70 to 80. The lamp life is about 12,000 hours, which is much more than the life of filament lamps.

Linear fluorescent lamps have evolved over time from old T12 to T8 and later to T5 lamps. The reduction in tube diameter and improvement in technology, including the use of electronic control gear (described later), have resulted in efficacy improvement from about 50 lm/W in the 1970s to more than 100 lm/W.

Compact fluorescent lamps.   Compact fluorescent lamps (Fig. 7.4) fold the discharge path and are therefore smaller in size compared to linear fluorescent lamps. The ballast is normally integrated with the lamp so that they can be easily fitted into standard sockets used by incandescent lamps. They have a much higher lamp efficacy compared to incandescent lamps and use only about 20 to 25 percent of electrical power (compared to incandescent lamps) to produce the same amount of light.

They also have a much longer life, which is about 8000 hours, compared to only 1000 hours for filament lamps. Although compact fluorescent lamps cost more than incandescent lamps, they are more economical to use due to their lower energy consumption and longer life.

They are available in a variety of shapes. The color temperature ranges from 2700 to 6000 K compared to 2700 to 3000 K for incandescent lamps. The CRI is about 82 to 88.

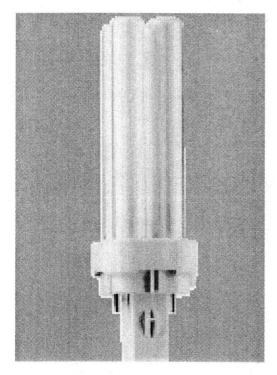

Figure 7.4 Compact fluorescent lamps. (*Courtesy of Philips.*)

7.3.4 High-intensity discharge lamps

Metal halide lamps, mercury vapor lamps, and sodium lamps are called high-intensity discharge (HID) lamps. They are able to provide better efficacy and longer life than fluorescent lamps. They are commonly used in high-bay lighting for industrial applications, outdoor floodlighting, and street lighting.

Mercury vapor lamps (Fig. 7.5) use a mercury arc tube and a filament in the same envelope coated with a fluorescent material. They have an efficacy of about 30 to 55 lm/W, which is lower than other HID lamps. They have a longer life than the other HID lamps and are therefore preferred for high-bay applications in areas where lamp replacement is difficult.

Metal halide lamps (Fig. 7.6) are the basic mercury vapor lamp with iodides of metals added to them. They have a higher efficacy than mercury lamps (about 95 lm/W) and produce a higher quality of light.

There are two types of sodium lamps, namely, high-pressure sodium lamps (Fig. 7.7) and low-pressure sodium lamps (Fig. 7.8). High-pressure sodium lamps have a ceramic arc tube with a clear outer envelope. They have a higher efficacy than metal halides and mercury vapor lamps. They also have better color rendering than low-pressure sodium lamps and are generally used for street lighting, industrial lighting, and floodlighting.

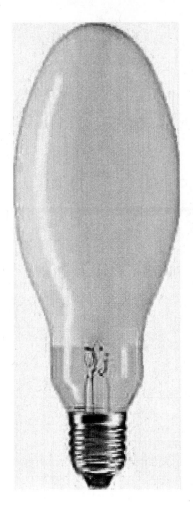

Figure 7.5   Mercury lamp. (*Courtesy of Philips.*)

Figure 7.6   Metal halide lamp. (*Courtesy of Philips.*)

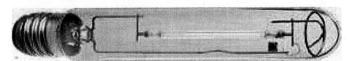

Figure 7.7   High pressure sodium lamp. (*Courtesy of Philips.*)

Figure 7.8   Low-pressure sodium lamp. (*Courtesy of Philips*)

Low-pressure sodium lamps consist of a U-tube containing the discharge and an outer thermal jacket. They have a high efficacy (over 150 lm/W) but produce light only in the yellow part of the spectrum. Therefore, these lamps are generally used only for outdoor applications such as street lighting and security lighting.

The main advantages of HID lamps are that they are compact sources that can produce large amounts of lumens and are available in a wide range of wattages. However, they have a warm-up time and therefore take a few minutes to reach the full light output. They also cannot be restarted for a few minutes after switching off as they need to cool down before starting.

7.4   Lighting Levels

The lighting level or lux level required for a space depends on the type of space, tasks performed in the space, and other visual requirements. General guidelines for the illuminance range for different applications that need to be used when designing of lighting systems are available in lighting reference books and codes of practice. A summary of recommended lighting levels for some common building spaces are given in Table 7.6.

The lighting levels given in Table 7.6 are used generally as a guideline to ensure that lighting levels provided are adequate for the specific

TABLE 7.6   Recommended Illuminance Levels

Type of area	Illuminance (lux)	Recommended design value (lux)
General offices, conference rooms, computer workstations	300–750	500
School classrooms	200–500	300
Shops, departmental stores	300–750	500
Supermarkets	500–1000	750
Hospitals	200–500	300
Lobbies, corridors	100–200	150
Hotel rooms:		
General	75–150	100
Local	200–500	300
Car parks:		
Parking areas	10–20	15
Entrance	50–300	100

tasks to be performed while preventing unnecessary wastage of electricity due to excessive lighting levels.

7.5   Lighting Power Density

The amount of electricity used for lighting also depends on the type of lighting used (efficacy) and other lighting design criteria such as the type of luminaries used, location and spacing of luminaries and whether reflectors or covers are used. The overall efficiency of a lighting design can be evaluated using the lighting power allowance (or lighting power density) which is computed in W/m^2 or W/ft^2.

ASHRAE Standard 90.1 prescribes two methods to determine compliance for interior lighting. The simplest method of the two is the "building area method" where the gross lighted floor area is multiplied by the allowable lighting power density in W/m^2 for a particular building area type to determine the maximum allowable installed lighting power.

For offices, the maximum allowed is 10.8 W/m^2 (1 W/ft^2) which can easily be achieved using T8 lamps with electronic ballast. As a guideline, the lighting power density for office spaces, shops and car parks should not exceed 10 W/m^2, 20 W/m^2, and 5 W/m^2, respectively.

7.6   Common Energy Saving Measures

Based on the lighting fundamentals and concepts explained earlier, some possible energy saving measures that can be applied for lighting systems are described next.

7.6.1   Reducing lighting levels

As explained earlier, the lighting level for a particular space depends on the tasks to be performed in the space and other visual requirements for the space. Generally, higher lighting levels lead to higher lighting energy consumption. Therefore, lighting levels should be minimized and maintained based on the recommended values in Table 7.6.

For new installations, this can be achieved through good lighting design by optimizing factors such as lamp wattage, number of lamps, and lamp spacing.

For existing installations where it is not cost effective to redesign lighting systems, other means such as delamping, use of task lighting, and replacement of lamps can be considered to reduce energy consumption.

Delamping.   Delamping involves removing one or more lamps from a fixture (that has more than one lamp) in areas where the lighting level is higher than required. Delamping could be done by simply removing lamps from fixtures or by converting existing luminaries to use less number of lamps, which may involve using conversion kits with reflectors and adaptors. The latter option, although more costly, is generally preferred as simply removing lamps could sometimes give an impression of the building being poorly maintained. Delamping can be carried out for areas such as offices, car parks, corridors, and toilets.

Use of task lighting.   In many building spaces general lighting is provided uniformly at a higher level so that the required lighting level can be achieved on the work surfaces (Fig. 7.9). However, since the quantity of light is inversely proportional to the square of the distance from the light source (inverse square law), a more intense lighting level is required for general lighting compared to task lighting to achieve the same lighting level at the required area.

In general lighting applications savings can also be achieved by locating the ceiling fixtures directly over the work areas.

Replacing lamps.   Another strategy is to replace high wattage lamps with lower wattage lamps. For example, T12 lamps can be replaced with T8 lamps and T8 lamps can be replaced with T5 lamps. Although the lumen levels for the higher efficacy replacement lamps may sometimes be lower, the impact on the general lighting level is normally not noticeable.

Less general lighting

Task lighting

Figure 7.9   Effective use of task lighting. (*Courtesy of Philips.*)

7.6.2   Use of energy-efficient lamps

The amount of light emitted by a lamp per unit of electrical power consumed is the luminous efficacy of a lamp. The units of luminous efficacy are lumens/watt (lm/W). The higher the efficacy, the better the efficiency of the lamp. Since different types of lamps have different efficacies, lamps with low efficacy can be replaced with those having higher efficacy.

Incandescent lamps.   Incandescent lamps have a low efficacy of about 10 to 15 lm/W and, therefore, are not energy efficient. They can be easily replaced with compact fluorescent lamps (CFL), which have a much higher efficacy of about 50 to 60 lm/W. In general, compact fluorescent lamps consume only about 20 to 25 percent of the power consumed by incandescent lamps to produce the same amount of light.

A comparison of power ratings for incandescent and compact fluorescent lamps for producing equivalent illuminance levels is given in Table 7.7.

Compact fluorescent lamps also have a life span of about eight times that of incandescent lamps (8000 hours compared to 1000 hours). Although compact fluorescent lamps are costlier than incandescent lamps, the extra cost can be normally paid back within about one to two years of operation due to the lower energy consumption and longer life span of these lamps.

Example 7.1   Calculate the simple payback period for replacing 100 nos. of 75 W incandescent lamps with 15 W compact fluorescent lamps if they operate 24 hours a day, seven days a week, and the electricity tariff is $0.10/kWh.

The lamp cost and lamp life are given below for the two types of lamps.

	Incandescent 75 W	Compact fluorescent 15 W
Unit cost ($)	1	15
Lamp life (hours)	1000	8000

Energy savings per lamp = (75 − 15) = 60 W

Energy savings per day = (60 × 100 × 24)/1000 kWh = 144 kWh/day

Annual energy savings = 144 × 365 × $0.10 = $5250

Payback period based purely on energy savings

$$= \$(15 \times 100)/\$5{,}250$$

$$= 0.3 \text{ years (3.4 months)}$$

TABLE 7.7   Comparison of Lamp Power for Incandescent and Compact Fluorescent Lamps

Incandescent lamp	40 W	60 W	75 W	100 W
CFL	9 W	11 W	15 W	20 W

Payback period taking into account longer lamp life:

$$\text{Annual cost of incandescent lamps} = \$1 \times (8760/1000) = \$8.87$$

$$\text{Annual cost of CFL} = \$15 \times (8760/8000) = \$16.43$$

$$\text{Total additional cost} = \$(16.43 - 8.87) \times 100 = \$756$$

$$\text{Payback period} = 756/5250 = 0.14 \text{ years (2 months)}$$

Linear fluorescent lamps. Linear fluorescent lamps are used for many general applications such as offices, corridors, and car parks. These lamps have an efficacy of 69 to 100 lm/W and are more energy efficient than compact fluorescent lamps.

In the past, the linear fluorescent lamps used were T12 (40 W) lamps (38-mm diameter), which have an efficacy of 69 lm/W. They have been gradually replaced by the T8 lamps, which have a smaller tube diameter (26 mm) and are more energy efficient (90 lm/W). Since T12 and T8 lamps have the same tube length, T12 lamps can be replaced with T8 lamps without having to replace the fixtures. Typically, a two lamp, 1200-mm long T12 luminaire with low-loss magnetic ballast will consume about 86 W compared to 72 W for an equivalent T8 system. Therefore, savings of about 14 W can be achieved in such applications.

Example 7.2 Estimate the daily electrical energy savings that can be achieved if a 1200 mm (4 feet) long two lamp luminaire using T12 lamps is converted to use T8 lamps. Take the operating hours to be 12 hours a day.

$$\text{Energy savings per luminaire} = (86 - 72) = 14 \text{ W}$$

$$\text{Energy savings per day} = (14 \times 12)/1000 \text{ kWh}$$

$$= 0.168 \text{ kWh/day}$$

The newer T5 lamps are even more compact (16 mm diameter) and have an efficacy of about 100 lm/W. As such, T5 lamps should be considered as an alternative to T8 lamps for applications such as offices and corridors. However, due to the shorter tube length of T5 lamps, they cannot be directly fitted to existing T8 fixtures. Therefore, use of T5 lamps should normally be considered for new installations rather than for retrofitting existing fixtures. However, T5 lamps with integrated adaptors designed to fit into existing T8 fixtures for retrofit applications are also currently available in the market.

High-bay. For high-bay lighting applications, such as warehouses, factories, large stores, and sports facilities, metal halide and high-pressure sodium lamps should be used instead of mercury vapor lamps as they have a higher efficacy. High-pressure sodium lamps are also able to produce a fuller spectrum of light, which can help improve nighttime visibility.

Similarly, for outdoor lighting, metal halides, high-pressure sodium and, where acceptable, low-pressure sodium lamps can be used. Low-pressure sodium lamps have an efficacy of about 175 lm/W, which is higher than that of other types of HID lamps. However, since the lighting quality of low-pressure sodium lamps is not good, they are generally used only for outdoor security lighting and street lighting.

Example 7.3   Calculate the simple payback period for replacing 100 nos. of 600-mm (2 feet) long T8 lamps with T5 lamps costing $30 each if they operate 24 hours a day, seven days a week, and the electricity tariff is $0.10/kWh.

The power consumption and lamp life are given below for the two types of lamps.

	T8	T5
Power consumption of lamp & ballast (W)	22	15
Lamp life (hours)	1000	8000

Energy savings per lamp = $(22 - 15) = 7$ W

Energy savings per day = $(7 \times 100 \times 24)/1000$ kWh = 16.8 kWh/day

Annual savings = $16.8 \times 365 \times \$0.1 = \$613.20$

Total cost = $\$30 \times 100 = \$3000$

Payback period = $3000/613.20 = 4.9$ years

7.6.3 High-efficiency electronic ballast

Ballasts are used to start and operate fluorescent and HID lamps. Ballasts provide the voltage necessary to strike the arc discharge and to regulate current drawn by the lamp to maintain light output.

The two main types of ballasts are magnetic and electronic. Conventional magnetic ballasts have a core of magnetic steel laminations surrounded by coil assemblies, while electronic ballasts have electronic components. These electronic ballasts (Fig. 7.10) operate at high frequencies, between 20,000 and 60,000 Hz, which makes power-to-light conversion more efficient than for magnetic coil ballasts. Electronic ballasts are superior to magnetic ballasts because they are typically 30 percent more energy efficient. There are also improved versions of the magnetic ballast called low-loss ballasts, which have lower losses due to the design and construction of the core and coil assembly.

Fluorescent lamp ballasts operate typically as preheat, rapid start, and instant start. In the preheat type operation, the lamp electrodes are heated prior to application of the starting voltage to strike the arc. The lamp flickers when starting in the preheat operation. In rapid start, the

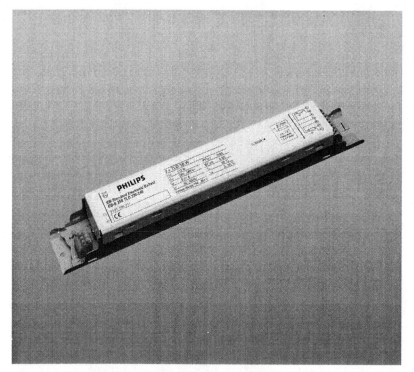

Figure 7.10   Typical electronic ballast. (*Courtesy of Philips.*)

electrodes are heated before starting and during operation. Usually, starting is smooth compared to the preheat operation. In instant start, lamp electrodes are not heated and the ballast provides a high open-circuit voltage across the electrodes to strike the arc. The lamp life tends to be shorter for this type of operation compared to rapid start.

Ballasts use energy during operation, which is called ballast losses. Ballast losses vary depending on the type of ballast used. Magnetic ballasts have the highest losses followed by low-loss and electronic ballasts. In applications using magnetic or low-loss ballasts, the actual input power is normally higher than the lamp rated watts. However, in applications using electronic ballasts, due to their lower losses, the total system input power can be less than the lamp rated power.

A comparison of power consumption for a fixture consisting of two 36 W lamps (1200 mm length) with different types of ballast is given in Table 7.8.

Example 7.4   Calculate the simple payback period for replacing 100 nos. of low loss ballast with electronic ballast for an application using 36 W lamps if they operate 24 hours a day, 7 days a week, and if the electricity tariff is $0.10/kWh. Take

TABLE 7.8   Comparison of Power Consumption
for a Fixture with Two 36 W Lamps

Ballast	Watts	Savings
Magnetic	84	–
Low loss	72	14%
Electronic	64	24%

the unit cost of electronic ballast to be \$20 (supply and installation) and the saving in power consumption to be 4 W per lamp.

$$\text{Energy savings per day} = (4 \times 100 \times 24)/1000 \text{ kWh} = 9.6 \text{ kWh/day}$$

$$\text{Annual energy savings} = 9.6 \times 365 \times \$0.10 = \$350.40$$

$$\text{Total cost of ballast} = \$20 \times 100 = \$2,000$$

$$\text{Payback period} = 2000/350.40 = 5.7 \text{ years}$$

7.6.4   Luminaires

Luminaires consist of lamps, ballast, lamp holders, optical devices, and a housing. Optical devices normally consist of components such as reflectors, lenses, or louvers to effectively deliver light to the space or work plane. The optical efficiency of a luminaire depends mainly on the material properties and geometry of the optical components.

Linear fluorescent lamps are the most commonly used lamps for general lighting in commercial and industrial applications. Due to the cylindrical shape of linear fluorescent lamps, if reflectors are not used, about half of the light emitted by them would be reflected on to the ceiling. Therefore, various types of reflectors are used to improve the optical efficiency of luminaries used in linear fluorescent lamps.

Most standard reflectors used in luminaries of fluorescent lamps have a reflectivity of about 70 to 80 percent However, reflectors made of special materials capable of improving the reflective effectiveness are available. These reflector materials have a mirror-like finish that allow better redirection of light rays.

Some of the special reflector materials are anodized specular aluminum, having a total reflectivity of 85 to 90 percent, and vacuum-deposited specular silver applied on a polyester surface, having a total reflectivity of 91 to 95 percent.

Fluorescent luminaires can be retrofitted with such new reflectors, often resulting in increased useable light, enabling building owners to remove lamps and save energy while still maintaining acceptable lighting levels. In some applications, reflectors can be used to retrofit existing luminaries to convert them from 4- to 2-lamp operation, thereby reducing the lighting energy consumption by about 50 percent.

Maintenance can also impact the optical efficiency of luminaries, and the light available in the space or at the working plane can be reduced due to the accumulation of dirt, discoloring of lenses and reflectors, and aging of lamps. Therefore, regular maintenance is required to maximize the lighting levels by cleaning dirty lamps and fixtures, replacing faded reflectors, lenses and lamps which are nearing the end of their useful life.

7.6.5 Lighting controls

Energy consumed by lighting can also be reduced by minimizing their usage by better matching operations with demand through lighting controls. Various systems such as timers, occupancy sensors, and light sensors can be used to control lighting operations.

Timer schedules. Simple timers can be used to switch on and off all or some lighting circuits at predetermined times based on occupancy schedules. Provision for manual override can be incorporated into the controls so that occupants can extend the operating hours of lighting circuits based on individual requirements. Lighting control systems can consist of simple timers that have 24-hour clocks to switch on and off lighting daily at preset times, or more sophisticated timers that can be used to program lighting schedules for a year or more, where holidays and other special requirements can be programmed in advance. Often lighting operating schedules can also be programmed into building automation systems to control the operating hours of lighting.

Occupancy sensing. Occupancy sensors can also be used to switch on lighting when a space is occupied and switch off the lighting after a preset time delay when the space is not occupied. Typical applications for occupancy sensors are in toilets, car parks, meeting rooms, storage areas, and common corridors.

The two basic technologies used in occupancy sensing devices are infrared and ultrasonic. Infrared sensors scan the area around them to detect heat generated by occupants. They are ideal for small open areas such as offices and classrooms. Ultrasonic sensors emit high frequency sound waves to detect occupancy. They are generally used in large or obstructed areas. Due to the relative advantages and disadvantages of the two types of technologies, sensors that incorporate both types of technologies are available with more effective sensing capabilities.

Light sensing. Making adequate use of natural light is another way to reduce a building's energy load (Fig. 7.11). Exterior and interior areas of buildings, which are exposed to natural light, can have light sensors to switch off or provide dim artificial lighting when sufficient natural light is available. The daylight controls can be the open-loop type where

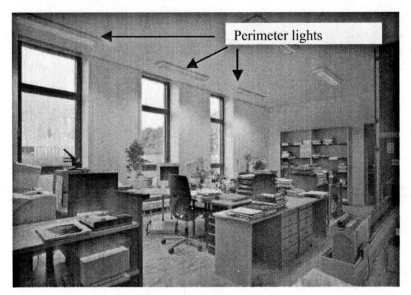

Figure 7.11   Effective use of natural light. (*Courtesy of Philips.*)

the sensor detects available daylight or the closed-loop type where the sensor detects available light at a work space.

7.6.6   Lighting energy saving devices

There are many *energy saving devices,* which help to reduce lighting energy consumption. The design of these equipment vary from one to the other but most contain transformers to reduce the voltage to a predetermined value after the lamps are switched on. This helps to reduce the power drawn by the lighting, but result in a drop in lighting level. These devices normally only work with lighting using magnetic or low-loss ballasts.

Since reduction in voltage leads to a reduction in the current drawn, the reduction in power consumption is proportional to the square of the reduction in voltage. The reduction in the illuminance level is approximately proportional to the reduction in voltage.

Table 7.9 shows some typical data collected from a trial carried out using a power saving device set at different voltage reduction levels.

TABLE 7.9   Results of Trial Carried Out with a Lighting Power Saving Device

Voltage reduction	17%	21%	29%
Power (kW) reduction	32%	37%	51%
Lighting level (lux) reduction	19%	22%	31%

Various tests have been conducted on such power saving devices, including under the National Lighting Product Information Program (NLPIP), which is sponsored by U.S. Environmental Protection Agency (EPA), and other state organisations. Based on the report, lighting power saving devices using transformers to reduce voltage are able to reduce lighting power while lowering light output. The report states that ballast life can improve due to lowering of ballast temperature by power saving devices and that there are no published reports documenting their effect on lamp life at lower than rated power.

Fluorescent-lamp life can be affected by the lamp's current crest factor (CCF) and the electrode starting temperature. Since power saving devices limit lamp current, they do not increase lamp CCF. As most power saving devices also provide the normal voltage during starting, they do not alter the starting temperature. Therefore, power saving devices are not expected to have a significant impact on lamp life if the voltage is not reduced significantly.

Since power saving devices normally need to be installed at the lighting DBs (distribution boards), they are generally used on lighting circuits that have many lamps on a single circuit. They are ideal for car park lighting systems because they tend to have all the lighting on a few circuits and also because they are able to accept the reduction in lighting levels associated with such devices. Care should be taken to ensure that nonlighting loads are removed from lighting circuits to prevent possible damage to equipment when operating at lower voltage.

7.6.7 Daylighting

Sunlight falls on the exterior surfaces of most buildings even on cloudy days. This natural light can be captured through daylighting techniques to illuminate interior spaces of buildings and help reduce the energy consumption of artificial lighting.

Hundreds of years ago when artificial light sources were not available, daylight was the only efficient light source available and one of the main goals of architecture was to have large openings to bring natural light into buildings. With the availability of efficient artificial lighting systems, capturing daylight has become less important in modern building architecture.

However, as artificial lighting accounts for a significant portion of electrical energy consumed in buildings, daylighting techniques can be incorporated into building architecture to make buildings more energy efficient.

Daylighting strategies incorporated into building designs depend on factors such as the availability of natural light (building latitude), climate, presence of obstructions, and building shape.

For example, high latitude areas have distinctive summer and winter climates with appreciable variation in seasonal daylight levels compared to lower latitudes where daylight levels do not have much seasonal variation. Therefore, in high latitudes when winter daylight level is low, the building design objective is to maximise daylight penetration as opposed to restricting daylight penetration into buildings at low latitudes.

In most situations, other structures in the surrounding area obstruct natural light. Sometimes, building features can also act as obstructions and the design needs to take into account such obstructions to maximise the use of daylight.

The shape of buildings also affect daylighting strategies. For instance, the daylighting needs of a particular building may be satisfied with conventional windows whereas another may require more complex design features.

Some common daylighting features used in buildings to maximise the use of natural light are windows, skylights, light shelves, and light tubes.

Windows.   Windows are the most commonly used daylighting feature in buildings. They are able to bring daylight into the perimeter areas of buildings. The amount of daylight transmitted depends on factors such as the window size, orientation, and material properties. To prevent direct solar radiation from being transmitted into buildings, selective window coatings, blinds, or other shading devices can be used.

Skylights.   Skylights are transparent areas of a roof, provided to allow daylight to enter a space. They can range from small windows on the roof to the entire roof of atrium areas. Skylights can be constructed using glass or special materials that can help diffuse the light entering a space while preventing transmission of direct sunlight, as shown in Fig. 7.12.

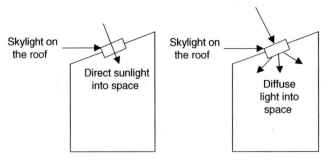

Figure 7.12   Skylights with normal and special materials.

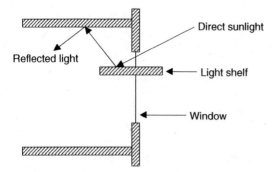

Figure 7.13 Arrangement of a typical light shelf.

Light shelves. A light shelf (Fig. 7.13) is a passive architectural feature that consists of a horizontal reflecting surface that permits daylight to enter into a building. Light shelves help to prevent direct sunlight from entering a space by reflecting sunlight onto the ceiling, thereby minimizing glare and increasing the lighting level in the space. They work best at high solar angles but can be extended inwards at low angles to capture the direct rays.

Light tubes. Light tubes and light pipes have internal light reflecting surfaces that help to transmit natural light into the interior spaces of buildings to minimise the use of artificial lighting. Figure 7.14 shows the arrangement of a typical light tube.

7.7 Summary

Typically, lighting accounts for more than 20 percent of the electrical energy consumption in commercial buildings. In air-conditioned buildings they also indirectly account for some of the power consumed by

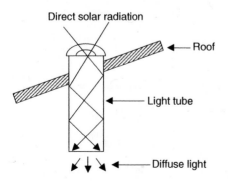

Figure 7.14 Arrangement of a light tube.

air-conditioning systems as the heat added by lighting has to be removed by building cooling systems. Therefore, lighting systems need to be designed to be efficient to ensure the overall energy efficiency of buildings.

The chapter provided a description of the types of lighting used in buildings with definitions of some important terms. Thereafter, various energy saving measures relating to building lighting systems, such as optimizing lighting levels, improving lighting system efficiency, lighting controls, and daylighting were described.

Review Examples

7.1. In an office space, lighting is provided by 150 nos. of T8 lamps on a single lighting circuit. Each luminaire has three lamps and uses a magnetic ballast. The lighting level measured at the working space is on an average 700 lux.

Describe three possible energy saving measures for this application.

7.2. Describe three lighting control strategies that can be used to minimize lighting energy consumption in buildings. For each strategy, list two possible applications.

7.3. Calculate the simple payback period for replacing 1000 nos. of 60-W incandescent lamps with 11-W compact fluorescent lamps, based on the following:

Unit cost of 11-W compact fluorescent lamp = $12

Unit cost of 60-W incandescent lamps = $1

Compact fluorescent-lamp life = 8000 hours

Incandescent-lamp life = 1000 hours

Operating hours = 12 h/day × 5 days a week

Electricity tariff = $0.10/kWh

7.4. Calculate the simple payback period for replacing 100 nos. of 1200-mm (4 feet) long T8 lamps with T5 lamps, based on the following:

Unit cost of T5 (lamp + fixture) = $40

Operating hours = 24 h/day × 7 days a week

Electricity tariff = $0.10/kWh

Power consumption of T8 lamp = 40 W (lamp + ballast)

Power consumption of T5 lamp = 28 W (lamp + ballast)

7.5. A particular luminaire has two 36-W lamps using a single magnetic ballast. Calculate the simple payback period for replacing the magnetic ballast with an electronic ballast, based on the following:

Unit cost of electronic ballast = $20 (supply and installation)

Operating hours = 24 h/day × 7 days a week

Electricity tariff = $0.10/kWh

Power consumption per luminaire with magnetic ballast = 84 W

Power consumption per luminaire with electronic ballast = 64 W

Building Electrical Systems

8.1 Introduction

Building electrical systems comprise of transformers, distribution systems, switchgear, control panels, and motors. Most of the electrical energy consumed in buildings is accounted for by motors, which are used for operating equipment such as the air-conditioning plant, pumps, elevators, and fans. Office equipment such as computers, printers, copiers, and lighting account for most of the balance electrical energy used in buildings.

Utility companies, which supply electricity, charge for electrical energy consumed in kWh. The tariff paid for each unit of kWh used can be fixed rate, tiered (peak, off-peak, shoulder, and so on), or based on the time of day. Depending on the tariff structure, utility companies also charge for the maximum power demand (kW) and power factor.

For totally resistive loads, electrical power is the product of voltage and current used.

$$\text{Power} = \text{voltage} \times \text{current} \qquad (8.1)$$

The basic unit of power is watt (W) and is normally measured in kilowatts (kW), which is 1000 watts. Voltage is measured in volts (V) and current in amperes (A).

Electrical energy is power used over a time and is the product of power and time.

$$\text{Electrical energy} = \text{power} \times \text{time} \qquad (8.2)$$

Therefore, electrical energy of 1 kWh is 1 kW used over one hour.

In circuits that have inductive elements such as motor windings, transformer windings, and fluorescent lamp ballasts, there are two components

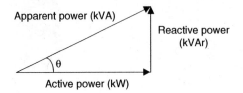

Figure 8.1 Vector diagram for power.

of power. One is the actual power absorbed by the component to do useful work, called real power (or active power), and the other is the reactive power used for magnetizing the magnetic elements. The apparent power is the vector sum of the reactive and active power and is normally computed in kVA (product of volts and amperes divided by 1000).

Power factor is the ratio of active power to the apparent power, as shown in Fig. 8.1. The power factor ranges from zero to 1.0. The highest power factor of 1.0 is achieved if there is no reactive power, as in the case of totally resistive loads.

$$\text{Power factor} = \cos\theta = \frac{\text{active power (kW)}}{\text{apparent power (kVA)}} \qquad (8.3)$$

Therefore, Power kW = kVA × cos θ

The power factor for a building depends on the different equipments and systems used in that building. In the case of motors, the power factor can also vary with the load applied on the motor. Utility companies usually penalize consumers if the overall power factor is below a certain value.

The different aspects of how electricity charges paid to utility companies can be lowered by reducing electrical energy usage (kWh), maximum power demand (kW), and improvement of power factor are described in the following sections of this chapter.

8.2 Efficiency of Motors

Efficiency is the measure of how well an electrical device converts the power consumed into useful work. Some devices like electric heaters can convert 100 percent of the power consumed into heat. However, in other devices such as motors, the total energy consumed cannot be converted to usable energy as a certain portion is lost and is not recoverable because it is expended in the losses associated with operating the device (Fig. 8.2). Therefore, it is necessary to provide more than 1 kW to produce 1 kW of mechanical output.

$$\text{Motor efficiency is } \eta = \frac{\text{power}_{\text{OUT}}}{\text{power}_{\text{IN}}} \qquad (8.4)$$

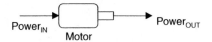

Figure 8.2 Definition of motor efficiency.

As shown in Fig. 8.3, the largest single loss in a motor is the stator resistance loss (Stator I^2R) followed by the rotor resistance loss (Rotor I^2R). These are followed by the core losses (hysteresis and eddy current) resulting from the cycling magnetic forces within the motor. Other motor losses are friction loss from bearings, windage loss due to drag during rotation and motor cooling, and stray losses.

The efficiency of motors depends on size, and normally ranges from about 78 to 93 percent for standard efficiency motors. In addition to these standard motors, some motor manufacturers also produce premium efficiency motors, which operate at efficiencies about 3 to 7 percent higher than the standard designs.

In these energy efficient motors, losses are reduced by:

- Use of wire with lower resistance
- Improved design of the rotor electric circuit
- Higher permeability in the magnetic circuits of the stator and rotor
- Use of thinner steel laminations in the magnetic circuits
- Improved shape of the steel stator core and rotor magnetic circuits
- Smaller gap between stator and rotor
- Internal fan, cooling fins, and cooling air passages designed to reduce the cooling power requirement
- Use of bearings with lower friction

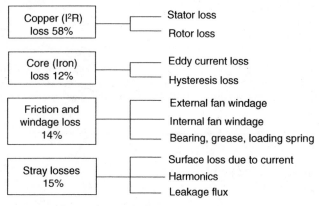

Figure 8.3 Losses in a typical motor.

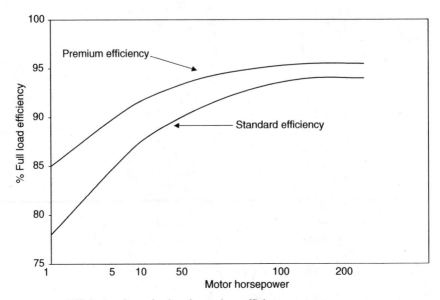

Figure 8.4   Efficiency of standard and premium efficiency motors.

Figure 8.4 shows the comparison of full-load operating efficiency for typical standard and premium efficiency motors. For motors over 7.5 kW (10 Hp), the improvement in efficiency possible from premium efficiency motors is 3 to 5 percent. Since the improvement achievable in efficiency is only a few percent, it is normally not economically viable to replace an existing motor with a premium efficiency motor merely based on savings. However, in case of new installations or motor replacements, premium efficiency motors should be considered in place of standard ones as the savings can normally pay for the incremental cost of the higher efficiency motors.

The following example illustrates how savings can be estimated for a case where a standard efficiency motor is to be replaced with a premium efficiency motor.

Example 8.1   Consider an 18.7 kW (25 Hp) motor functioning at 88 percent efficiency (standard efficiency), operating 24 hours a day 365 days a year. Calculate the savings that will result if it is replaced with a (premium efficiency) motor functioning at 93 percent efficiency.

Using Eq. (8.4) for the motor, $\text{Power}_{\text{IN}} = \dfrac{\text{power}_{\text{OUT}}}{\eta}$

Therefore, Energy saved in kWh = motor kW

$$\times \text{ operating hours} \times \left[\frac{1}{\eta_S} - \frac{1}{\eta_P} \right]$$

where η_S = efficiency of standard motor and η_P = efficiency of premium motor

$$\text{Annual electrical saving} = 18.7 \times 24 \times 365 \left[\frac{1}{0.88} - \frac{1}{0.93} \right] \text{kWh/year}$$

$$= 10{,}008 \text{ kWh/year}$$

In addition to the full load efficiency of motors, the operating efficiency of motors also depends on their loading. Figure 8.5 shows a typical motor characteristic curve, which relates the motor efficiency to its loading. As the figure shows, motor efficiency is close to its full load efficiency when loaded above 40 percent, but drops significantly if the motor is loaded lower than this value. Therefore, if a motor is loaded to less than about 40 percent, it would be a good candidate for replacement with a correctly-sized motor. However, care should be taken to ensure that the replacement motor can meet the starting torque required for the particular application, as in some instances motors are oversized to overcome a high starting torque.

Example 8.2 The rated efficiency of a standard 55 kW motor used for a water pump is 90 percent. When the pump is operating at its full load condition the power drawn by the motor is measured to be only 9 kW. Using Fig. 8.5, estimate the actual motor operating efficiency. If the pump operates 24 hours a day, estimate

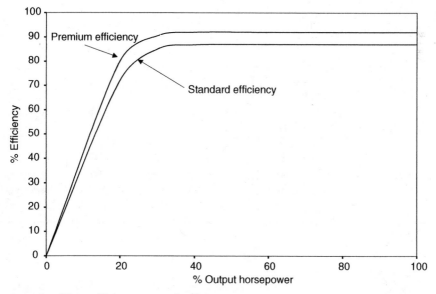

Figure 8.5 Motor efficiency versus loading.

the amount of electrical energy that can be saved if the pump motor is replaced with a smaller capacity motor.

The estimated loading of the motor is 9/55 = 16 percent

From Fig. 8.5, the motor operating efficiency is expected to be about 5 percent.

Based on this motor efficiency, motor output power = 0.55 × 9 = 5 kW

If a 5.5 kW motor is used for this application, the expected motor efficiency (from Fig. 8.4) is about 85 percent.

The input power for the new motor = 5.5/0.85 = 6.5 kW

Therefore, electrical energy saved = (9 – 6.5) × 24 = 60 kWh/day.

8.3   Variable Speed Drives

Many building systems are designed to operate at maximum load conditions. However, most building systems operate at their full load only for short periods of time. This often results in many systems operating inefficiently during long periods of time. Most such inefficient operations in buildings are encountered in air-conditioning systems that are normally sized to meet peak load conditions, which are experienced only for short periods of the day.

Some examples of such operations are:

- Chilled water and hot water distribution pumps
- Cooling tower fans
- Air handling unit fans
- Ventilation fans

As explained in earlier chapters, the efficiency of such systems can be improved by varying their capacity to match actual load requirements. As all the above are variable torque applications, the power required (to drive the pumps or fans) varies to the cube of the speed and, therefore, large power reductions result from small reductions in speed. The most common method is to modulate the speed of the motors of pumps and fans to vary their capacity using variable speed drives (VSDs).

The most commonly used motor in buildings today is the three-phase, asynchronous AC (alternating current) motor, which is both inexpensive and of very reliable construction. Ordinary asynchronous AC motors are designed to make the operating speed dependent on the frequency and the voltage connected to the motor. For example, when the mains supply is 400 V and 50 Hz, a four-poled asynchronous AC motor operates at an approximate speed of 1500 rpm.

A VSD [sometimes called a variable frequency drive (VFD)] is an electronic unit that provides infinitely variable control over the speed of

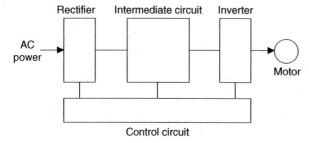

Figure 8.6   Configuration of a typical VSD.

three-phase AC motors by converting fixed mains voltage and frequency into variable quantities. It has no moving parts and uses a rectifier that is connected to the mains supply to generate a pulsating DC voltage and direct current, which is then passed to an inverter to generate the frequency of the motor voltage.

The configuration of a typical VSD is shown in Fig. 8.6. The main components of a VSD are the rectifier, inverter, intermediate circuit, and control circuit. The main functions of each component are listed here.

Rectifier.   Converts the AC to DC (direct current).

Intermediate circuit.   Stabilizes or smoothens the pulsating DC voltage and reduces the feedback of harmonics to the mains supply.

Inverter.   Converts DC voltage back into variable AC voltage with a variable frequency.

Control circuit.   Controls the VSD, enables exchange of data between VSD and peripherals, gathers and reports fault messages and carries out protective functions of the VSD.

8.4   Power Optimizing Devices

Various power optimizing devices, which are essentially "black boxes," are available in the market. They generally contain electronic circuits that monitor parameters such as motor load and power factor and continuously adjust the power supply to the motor to minimize consumption.

Tests carried out on some devices have shown that they are able to save on motor power consumption on variable torque applications, such as fans, pumps, and escalators. Comparison of VSDs with such devices indicate that the former is able to provide much higher savings for applications, such as pumps and fans, where the speed can be varied to match load conditions.

Tests with constant air volume (CAV) AHUs, where fan speed cannot be modulated, showed that power optimizing devices are able to achieve energy savings. The actual savings achieved depended on the motor loading. For motors loaded to only 60 to 70 percent of the rated capacity, energy savings of over 20 pecent was achieved.

For escalators, tests showed that power savings of about 10 percent is achievable. Since for escalators the actual motor power consumption is low (average 3 kW to 4 kW), the value of savings achieved may not financially justify the installation of such devices.

8.5 Transformer Losses

Transformers are equipment used to change the voltage of alternating current supplies. Normally, buildings are supplied electricity from the grid at high voltages such as 66 kV and 33 kV. The supply voltage, therefore, has to be reduced so that equipment in the building can make use of the electricity supplied.

Transformers consist of a primary winding and a secondary winding (Fig. 8.7). The primary winding is connected to the power source while the secondary winding is connected to the load. Electrical energy is transferred by induction and the ratio of primary voltage to secondary voltage is proportional to the number of turns of primary winding to the number of turns of secondary winding. Therefore, transformers can be step-down or step-up depending on whether the secondary voltage is lower or higher than the primary voltage.

There are two basic types of transformers, namely, dry type and liquid filled. In dry type transformers cooling is achieved through the free movement of air, while liquid filled transformers use a liquid to act as coolant and as an insulation dielectric.

Transformers are generally efficient and have low energy losses. The losses are normally about 1 to 2 percent of transformer capacity and depend on transformer type and size. Transformer losses are mainly due to copper losses, eddy current losses, and hysteresis losses.

These transformer losses can also be categorized into core losses (no-load losses) and coil losses (load losses), as shown in Fig. 8.8. The core losses originate in the steel core of the transformer, caused by the

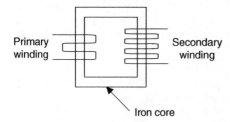

Figure 8.7 Simplified arrangement of a transformer.

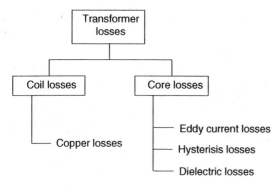

Figure 8.8 Transformer losses.

magnetizing current needed to energize the core. They are constant, irrespective of the load on the transformer. They continue to waste energy as long as the transformer is energized. Coil losses originate in the primary and secondary coils of transformers due to the resistance of winding materials.

Coil losses are due to power dissipated in the form of heat caused by the resistance of the conductor. The amount of power dissipated is directly proportional to the resistance of the conductor and the square of current flowing through it (I^2R).

No-load losses due to the magnetizing current, used to energize the core of the transformer, can generally be categorized into hysteresis losses, eddy current losses, I^2R losses due to no-load current, stray eddy current losses, and dielectric losses. Hysteresis losses and eddy current losses contribute the most and account for about 99 percent of the no-load losses, while stray eddy current losses, dielectric losses, and I^2R losses due to no-load current account for the rest.

Hysteresis losses come from the molecules in the core laminations resisting being magnetized and demagnetized by the alternating magnetic field. This resistance by the molecules causes friction, which results in heat. The Greek word, *hysteresis*, means "to lag" and refers to the fact that the magnetic flux lags behind the magnetic force. Choice of size and type of core material reduces hysteresis losses.

Eddy current losses occur in the core of the transformer due to the fluctuating magnetic field and induced voltage, which cause random currents to flow through the core dissipating power in the form of heat. Hysteresis losses are losses associated with magnetic domains of the core material.

Transformers are often sized based on expected demand, which far exceed actual load. Since the no-load loss is a function of the kVA capacity of the transformer and continue as long as the transformer is energized, transformers should be selected to better match actual load requirements. In situations where transformer capacity exceeds demand but the transformers have already been selected and installed, it may be possible to de-energize some transformers to minimize no-load losses. The no-load losses for transformers over 500 kVA is estimated to be about 0.3 percent of the rated capacity.

Example 8.3   A building uses one 750 kVA and three 1000 kVA capacity transformers. If the maximum total building load is measured to be 2500 kVA, estimate the electrical energy savings that can be achieved if the 750 kVA transformer is de-energized. Assume, the average power factor to be 0.9.

If the no-load losses are taken to be 0.3 percent, the savings by de-energizing the 750 kVA transformer

$$= \text{no-load kVA losses} \times \text{power factor} \times \text{operating hours}$$

$$= (750 \times 0.3\%) \times 0.9 \times 24 \times 365 = 17{,}739 \text{ kWh/year}$$

8.6   Elevators

Elevators can account for a significant portion of the electrical energy consumption in high-rise commercial buildings. In air-conditioned buildings, elevators typically account for more than 5 percent of the total consumption. The two main types of elevators are hydraulic type, which have hydraulic systems to provide movement, and traction type elevators, which use wire ropes pulled over sheaves driven by a motor. Traction elevators have counterweights linked to the elevator cab by a pulley system so that the counterweight lowers when the elevator cab rises and vice versa. This helps to reduce the weight to be lifted.

In elevators, energy is primarily consumed by the elevator motor, brake system, lights, and ventilation fans. The elevator motor generally consumes the most amount of power. However, under certain operating conditions the elevator motors can operate in regenerative mode, such as when the weight of the elevator cab and passengers is less than the weight of the counterweight when traveling up.

The electrical energy consumption of elevators depend on factors such as the type of motor drive used, the number of starts (door openings), carrying capacity, building height, and building occupancy. Data available from studies of elevators used in mid- and high-rise buildings indicate that average consumption of elevators range from about 5 to 40 kWh/day.

Some energy saving measures for elevator systems are listed below.

Type of elevator.   Hydraulic elevators are sometimes used in low-rise buildings due to lower cost. However, they are less efficient than traction elevators and consume about three times the amount of energy consumed by traction elevators for the same application. Therefore, for new installations in low-rise buildings and when replacing hydraulic elevators, traction elevators should be considered instead of hydraulic ones.

Type of drive. Old inefficient traction elevators use motor-generators as DC power sources compared to the later more efficient units that use solid state variable voltage (VV) and variable frequency (VF) drives with permanent magnet motors instead of induction units. The most modern systems, which convert line power to variable voltage, variable frequency (VVVF) to suit load and speed, are more energy efficient and should be considered for new installations and retrofits.

Regeneration. Traction elevators have counterweights that weigh about the equivalent of the weight of the elevator cab and half of its maximum load to help reduce the weight to be lifted by the elevator motor. Therefore, an empty elevator needs energy to descend (to overcome the counterweight) while a full elevator needs energy to ascend.

Similarly, an empty elevator ascending or a full elevator descending has potential energy that needs to be dissipated. In older elevators, this potential energy is dissipated in the form of heat in resistor banks. However, newer elevators are able to feed this regenerative power back into the building electricity distribution system, which helps to reduce the amount of electricity drawn from the mains supply. The amount of energy savings that can be achieved with regenerative systems depends on many factors but, typically, they are able to save as much as 30 percent compared to a geared traction system.

Controls. Modern elevator control systems can range from simple programs to schedule the turning on or off of all or some elevators during low usage periods to sophisticated systems that can "learn" from operations to position at specific locations based on usage patterns and the time of day. This helps to reduce waiting time for users and the distance traveled by elevators (to reduce energy consumption). Some sophisticated systems are even able to use control algorithms to optimize energy usage by considering the potential energy available from the counterweights of the different cabs based on their location. The savings achievable by using advanced control systems is estimated to be about 5 percent.

Lights and ventilation fans. One of the most basic energy saving measures is to program the control system to switch off lights and ventilation fans of elevators when they are not in use.

Another energy saving strategy is to use more efficient lighting systems for elevator cabs. Typically, elevator cabs use halogen and incandescent lamps, which result in a lighting power density of about 50 W/m^2 (compared to about 10 W/m^2 for typical office areas). As an alternative, compact fluorescent lamps can be used for this application and the power density can be lowered to between 10 and 15 W/m^2.

Reducing capacity.   In some buildings, due to reasons such as change in occupancy, the capacity of lifts can be much higher than necessary. For example, if a building initially designed for industrial use with elevators of capacity 10 tons each is converted for normal office use, the capacity of the elevators need to be only about 1 ton each. In such a situation, the elevators can result in much energy wastage due to the need for moving a cab that is much larger (heavier) than required and the need to overcome the extra weight of the counterweight designed for the original 10 ton capacity.

Energy savings can be achieved in such cases by replacing the elevator with a smaller capacity one or by reducing the counterweight (and capacity of the elevator).

8.7   Maximum Demand Reduction

Maximum demand for a building is the maximum power drawn from the grid in kW. The electrical demand is usually calculated by averaging the integrated power demand over a fixed interval (normally 30 minutes) using a maximum demand meter (Fig. 8.9). This computation is performed continuously for the same fixed time interval. If the maximum demand during a particular time period is lower than the previous value, the meter retains the previous reading. However, if the new reading is higher than the previously recorded highest maximum demand, then the new value is retained. The maximum value remaining at the end of a month is taken as the maximum demand for the building for that month.

Utility companies charge consumers for the maximum demand as their power generating equipment, distribution cabling, and switchgear need to be sized to satisfy the maximum demand requirements of the end users. For example, if a consumer has high power demand for a short

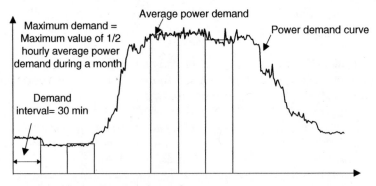

Figure 8.9 Maximum power demand.

period of time during the day compared to the demand during other periods of time, the utility company would still need to invest in additional infrastructure to meet this demand even though it is only required for a short period of time. Therefore, to compensate for this and to encourage consumers to reduce power demand, utility companies charge based on maximum demand.

Maximum demand charges paid depend on the actual tariff structure and can represent a significant portion of the utility bill. Therefore, considerable savings can be achieved by reducing the maximum power demand.

8.7.1 Load factor

The load factor is the ratio of average load to peak load during a specific period of time, expressed as a percentage.

$$\text{Load factor} = \frac{\text{average power demand (kW)}}{\text{maximum power demand (kW)}}$$

$$= \frac{\text{kWh for month}}{\text{maximum demand for month} \times 24 \times \text{no.of days in month}} \quad (8.5)$$

The load factor indicates to what degree energy has been consumed compared to the maximum demand or the utilization of units relative to total system consumption. The highest achievable load factor is 100 percent. Normally, the load factor for most buildings range from about 50 to 70 percent. Typical office buildings and retail malls, which operate about 10 to 12 hours a day, have a load factor of 50 percent, while hotels and industrial establishments, which operate 24 hours a day, normally have load factors over 70 percent.

Although the load factor depends on the operating characteristics of systems and the operational requirements, if the actual load factor is much lower than these approximate values, it is generally a good indicator of potential for reducing maximum demand.

The maximum demand for buildings can be reduced in different ways depending on the load characteristics of the building. They can be normally categorized into three demand management strategies—peak shaving, load shifting, and energy management—as illustrated in Fig. 8.10.

8.7.2 Peak shaving

Peak shaving (sometimes called peak clipping) involves reducing the maximum demand of a building. This strategy involves switching off

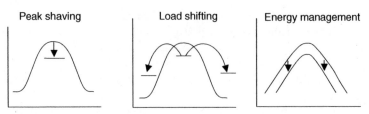

Figure 8.10   Demand management strategies.

equipment and systems that are considered to be nonessential during the period that a building experiences maximum power demand. The loads switched off are those such as ventilation fans, CAV AHU fans, and lighting which does not cause an increase in demand once they are switched on. Peak shedding can be achieved by the building's energy management system (EMS) by programming it to switch off equipment progressively, based on preset priority, as the building approaches its maximum demand. In addition to the priority set on the EMS for each of the equipment that can be switched off, other criteria such as the maximum duration (for switching off) and the maximum number of times they can be switched off can also be programmed on the EMS to avoid discomfort to occupants and to minimize wear and tear on equipment.

8.7.3   Load shifting

This load management strategy is similar to peak shaving as it involves switching off loads during peak period. However, the main difference in load shifting is that when it switches off equipment during peak periods, it is shifting this load to either before or after the peak demand period. Some examples can be switching off fans of VAV AHUs or turning off a chiller during the peak demand period, which then have to "work harder" when they are switched on later to meet the building's cooling requirements.

Another possible load shifting strategy is to use a thermal storage system for cooling, which can store either chilled water or ice produced during nonpeak periods and using it to cool the building during periods of peak demand. This enables either switching off all chillers or some chillers during peak periods to reduce maximum demand.

Other opportunities for load shifting also exist in situations where intermittent loads are encountered. An example of how maximum demand can be reduced by controlling the operation of loads that normally operate intermittently is illustrated in Fig. 8.11.

Example 8.4   An office building experiences a maximum power demand of 1200 kW in the mornings between 8.00 and 8.30 a.m. due to the switching on of the central air-conditioning system at 8.00 a.m. Thereafter, the power demand for the

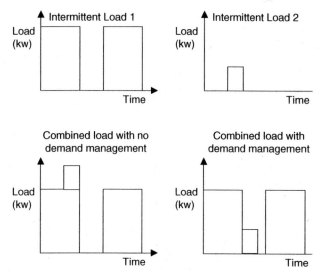

Figure 8.11 Illustration of demand reduction for intermittent loads.

entire day remains less than 900 kW. The total average electrical energy consumption for the building is 9000 kWh a day.

A trial carried out by the building management personnel shows that the building's maximum power demand can be maintained below 900 kW throughout the day if the central air-conditioning system is switched on earlier, at 7.30 a.m., with one less chiller operated over a longer period to cool the building. However, the total electrical consumption for the building increases to 9300 kWh a day.

Estimate the annual cost savings that can be achieved by switching on the air-conditioning system at 7.30 a.m. if the building operates 250 days of the year. Take the electricity tariff to be $0.10/kWh and the demand charge to be $1/kW (per month).

Saving in demand charges = (1200 − 900) × $10 = $3000/month = $36,000/year

Extra cost for consumption = (9300 − 9000) × $0.10 = $30/day

$$= \$30 \times 250 = \$7500/\text{year}$$

Net cost savings = (36,000 − 7500) = $28,500/year

8.74   Energy management

Energy management involves energy conservation and improving energy efficiency of a facility to reduce the power demand and energy consumption. As Fig. 8.6 shows this strategy helps not only to reduce the maximum demand but also to reduce the power demand at all times. As described in the earlier chapters, this is achieved by implementing various energy management and energy efficiency strategies.

8.8 Power Factor Correction

As explained earlier, power factor is the ratio of active power to the apparent power. The power factor ranges from zero to 1.0. The highest power factor of 1.0 is achieved if there is no reactive power, as in the case of totally resistive loads.

Electrical loads in commercial buildings and industrial facilities are not totally resistive and the reactive power component can be significant. Although only real power is consumed, the utility company has to make available the consumer the total power requirement made-up of both the real power and reactive power. Since reactive power constitutes an extra load on the power transmission and distribution system, utility companies penalize consumers if their power factor is low.

For utility companies this leads to extra load on the power transmission and distribution systems, leading to the need for higher capacity power plants, transmission cables, and switchgear. For consumers, low power factor results in overloading of equipment and higher energy losses due to higher current flow. Therefore, power factor should be as close as possible to unity and, generally, values above 0.9 are considered to be good.

Power factor can be improved by installing capacitors in parallel to reduce the reactive power. The power factor correction can be *static correction*, where capacitors are connected at each starter (Fig. 8.12) or *bulk correction*, where capacitors are connected at the distribution boards (Fig. 8.13). The effect of adding capacitors to reduce the power factor is illustrated in Fig. 8.14.

$$\text{Since, Power factor } = \frac{\text{active power (kW)}}{\text{apparent power (kVA)}} = \cos\theta \qquad (8.6)$$

and $\theta_2 < \theta_1$,

$\cos\theta_2 > \cos\theta_1$ (power factor is higher).

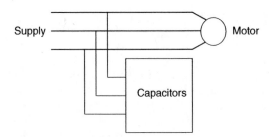

Figure 8.12 Static power factor correction.

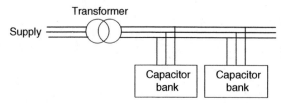

Figure 8.13   Bulk power factor correction.

Example 8.5   Consider a circuit that has active power of 1.5 kW, line current of 10 A, and line voltage of 240 V. Calculate the kVAr rating of a suitable capacitor that should be added to increase the power factor to 0.9.
  Before (refer to Fig. 8.15):

$$\text{Apparent power} = 240 \times 10 \text{ VA} = 2.4 \text{ kVA}$$

$$\text{Power factor, } \cos \theta_1 = 1.5/2.4 = 0.625$$

$$\theta_1 = 51.3°$$

$$\text{Reactive power} = \tan(51.3°) \times 1.5 = 1.87 \text{ kVAr}$$

After (refer to Fig. 8.15):
From Eq. (8.6), if power factor is to be 0.9,

$$\text{Apparent power} = 1.5 \text{ kW}/0.9 = 1.67 \text{ kVA}$$

$$\theta_2 = \cos^{-1}(0.9) = 25.8°$$

$$\text{Reactive power} = \tan(25.8°) \times 1.5 = 0.725 \text{ kVAr}$$

  Therefore, size of capacitor required = $1.87 - 0.725 = 1.15$ kVAr (capacitive)
  Since the line voltage is the same, the new current of 7 A (1.67/0.24) represents a reduction of 30 percent (from 10 A).

8.9   Equipment Standby Losses

Over the last 10 years, power consumed in commercial buildings has been rising due to the increased use of office appliances such as computers, printers, and copying machines. Today's commercial offices provide

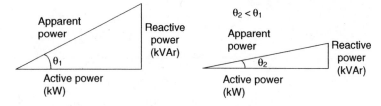

Figure 8.14   Effect of installing capacitors to improve power factor.

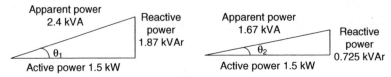

Figure 8.15   Vector diagram for Example 8.5.

a PC or workstation for almost all staff, many with 17" or 21" monitors. Further, due to increased cost of office space, offices are being compressed, leading to higher density of staff per unit floor area and higher power consumption in office buildings.

Most office equipment are actually used only for a few hours a day, leaving them switched on in standby mode for rest of the day while staff take breaks, eat lunch, attend meetings, and leave the office for other appointments. Further, printers, copy machines, and even computers are often left switched on after office hours.

Research has shown that such equipment can end up consuming up to 50 percent of normal power consumption even when in standby or idle mode if they are not switched off. In 1994, the Environmental Protection Agency (EPA) started the Energy Star program for computers to address the problem of wasted energy from computers and laser printers when they are left on and not being used. The program required that computer monitors have the ability to power down to power levels of 30 W or less. Laser printers were required to power down when idle to power levels of 30 to 60 W, depending on printing speed. A similar program was later also introduced for copying machines and fax machines.

Staff should therefore be encouraged to select and purchase office equipment complete with power management or energy saving features, that power down unnecessary components within the equipment, while maintaining essential functions or memory when the equipment is idle or after a user-specified period of inactivity. It should also be ensured that these energy saving features are not disabled by users (commonly encountered) as otherwise it would defeat the purpose of having them.

Steps should also be taken, where practically possible, to ensure that office equipment are switched off after office hours so that power consumption can be completely eliminated during such periods.

8.10   Summary

Motors used for operating equipment such as air-conditioning chillers, pumps, elevators and fans account for most of the electrical energy consumed in buildings, while computers, office equipment, and lighting account for the rest. Utility companies normally charge for electricity use based on energy consumption, power demand, and power factor. Various strategies to reduce these electricity charges were described in this chapter.

Review Questions

8.1. Compute the annual energy savings (in kWh) achievable for a 55 kW capacity motor if a 95 percent efficiency motor is used instead of an 91 percent efficiency motor. Take the operating hours to be 24 hours a day and 250 days a year.

8.2. Two 1500-kVA transformers are to be used in a building that normally operates 12 hours a day and 250 days a year. Calculate the kWh savings that can be achieved if 99.5 percent efficiency transformers are used instead of 98.5 percent efficiency transformers for this application. Assume the power factor to be 0.9 and the no-load losses to be the same for both types of transformers.

8.3. List five possible applications of variable speed drives (VSDs) in buildings to reduce electrical energy consumption.

8.4. List five possible energy saving retrofit strategies used for old elevators.

8.5. A building has an active power demand of 1500 kW and a power factor of 0.75. Calculate the kVAr rating of bulk correction capacitors required to be installed if the power factor is to be improved to 0.9.

Building Automation Systems

9.1 Introduction

Building automation systems (BAS) are computer based systems used for monitoring, controlling, and managing equipment and systems in buildings. Building systems normally managed by a BAS include air-conditioning, ventilation, lighting, heating, fire protection, electrical, and security systems.

Due to advances in information technology, a BAS can be integrated with other building services such as office automation systems, facility booking systems, and utility metering and billing systems. Therefore, in addition to monitoring and controlling, a BAS can assist in facility management.

The main components of a BAS include, sensors, actuators, controllers, data communication network, host computer, and software. Typical configuration of a BAS is shown in Fig. 9.1.

Building automation systems are a very important energy management tool. They can be used to control various energy consuming systems in buildings and perform many other functions required to optimize their operations. The following sections of the chapter provide a brief description of some of these functions and control strategies and how they can be used to optimize the operation of building systems.

9.2 Scheduling

Scheduling allows a BAS to take automatic action such as starting or stopping of equipment. The schedules are normally based on a one year calendar and can account for different operations on weekdays, Saturdays, Sundays, holidays, and different seasons.

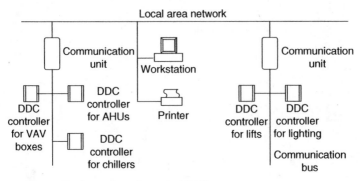

Figure 9.1 Configuration of a typical BAS.

This feature can be used to minimize energy consumption by operating equipment only when needed. For example, starting of chillers or boilers and AHUs can be staggered in the mornings to match actual requirements of the occupants in the different areas rather than switching them on all at the same time. Similarly, the scheduling feature can be used to start some equipment earlier on certain days, such as on Mondays, to account for the higher load or to switch off equipment during low occupancy periods, such as at lunch time.

The scheduling feature is much easier to use than manual timers as operating schedules can be easily entered, changed, or overridden remotely through a central workstation. It is also more flexible than normal timers since it can accommodate many different schedules to suit varying operating requirements.

9.3   Equipment Interlocks

BAS have interlock features where operation of groups of equipment can be interlocked so that they can be switched on or switched off together. This feature is useful to prevent accidental operation of equipment, which can cause harm or damage if they are not operated simultaneously with other equipment. For example, chilled water pumps, condenser water pumps, and cooling towers are usually interlocked with chiller operations. Similarly, boiler controls can be interlocked with the air intake and feedwater pumps for safety reasons.

This interlocking feature can also help to save energy by preventing operation of equipment when their associated equipment are switched off. A typical example is the interlocking of exhaust and supply fans to prevent one fan operating without the other, which could lead to infiltration or exfiltration of air. Similarly, the operation of kitchen exhaust fans can be interlocked to operate only when cooking is done.

9.4 Demand Limiting

Most BAS have a demand limiting algorithm, which can help to maintain the maximum power demand for a building below a set target. This feature helps to minimize maximum demand charges paid to utility companies.

Figure 9.2 illustrates a typical demand limiting feature, which predicts the future power demand based on the actual rate of increase of building demand and switches off (sheds) loads to ensure that the set demand limit is not exceeded. Similarly, the feature also normally allows the switching on (restoring) of loads when the demand drops.

In general, the demand limiting algorithm calculates the amount of load to be shed or restored and then sheds loads up to the cumulative load required to be shed. If there is an option to shed more than one load, the algorithm will shed loads based on set priority, where loads with lower priority will be shed before shedding loads with higher priority. If there are multiple loads with the same priority, the load that has been shed the least amount of time will be shed first. In this feature, other parameters, such as maximum and minimum off-time for each load, can be set based on operational requirements.

9.5 Duty Cycling

A duty cycling algorithm normally turns off loads for a short duration of time based on set priority and maximum off-times. The loads shed are "expendable" loads, which are loads that result in the reduction of both instantaneous power demand and energy consumption when shed.

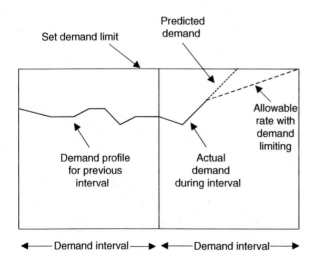

Figure 9.2   Illustration of a demand limiting strategy.

These loads cannot be "deferrable" loads as they would only result in instantaneous demand savings, because when they are restored the loads will work harder to make up for the time they have been off, resulting little or no energy savings.

Some expendable loads are fans of CAV AHUs and constant speed ventilation fans.

9.6 Trend Logging

Trend logging is a feature that allows recording of selected parameters at preset intervals of time. The feature can be used to log the value of parameters such as temperatures, flow rates, electrical power, and cooling demand. The recording interval can be set from one minute to about two hours.

The trend data can be used for various functions, ranging from trouble shooting to identifying energy saving opportunities.

Figures 9.3 to 9.6 show some typical trend data that can be extracted from a BAS. Figure 9.3 shows the power demand for a building and indicates that the maximum power demand occurs in the morning between 8 a.m. and 10 a.m. Meanwhile, Fig. 9.4, which shows the cooling load of the building, indicates that the maximum cooling load coincides with the maximum power demand for the building. This data can be utilized to reduce the maximum power demand for the building; for instance, by starting the chiller plant earlier to gradually cool the building.

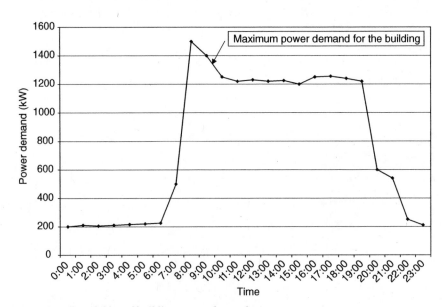

Figure 9.3 Trend data of building power demand.

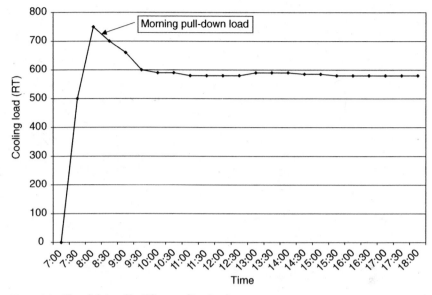

Figure 9.4 Trend data of building cooling load.

Another example of the usefulness of the trend logging feature in BAS is illustrated in Figs. 9.5 and 9.6. Figure 9.5 shows the trend data for chilled water supply temperature, which indicates that it exceeds the set value of 7°C after 12 noon, while Fig. 9.6 shows that the condenser water supply

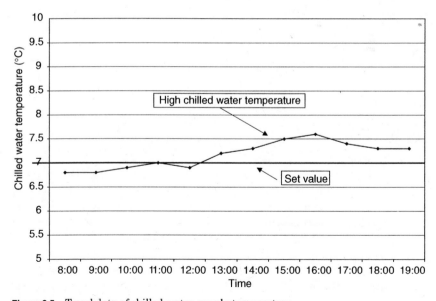

Figure 9.5 Trend data of chilled water supply temperature.

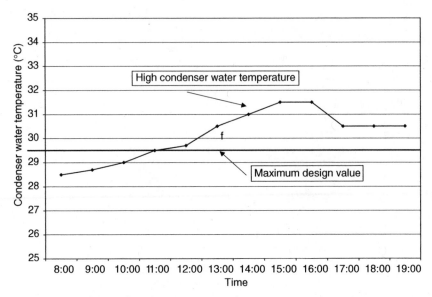

Figure 9.6 Trend data of condenser water supply temperature.

temperature exceeds the maximum design value after 11 a.m., indicating that the high condenser water temperature may be the cause of the chillers not being able to meet the chilled water temperature set point.

The trend logging feature can also be used for trouble shooting by trending multiple parameters. For example, if a chiller frequently trips, trending of related operating parameters, such as condenser water and chilled water flows and temperatures, may help to identify the cause as low flow or high condenser water temperature.

9.7 Alarms

This feature is designed to provide an alarm based on preset criteria. The alarm can be on the workstation to alert the operator or sent to a printer or other device, such as a pager.

The feature can be set to provide an alarm when a parameter reaches a certain condition such as a minimum or maximum value. It is very useful in energy management since it can alert the operator if a system deviates from a set condition.

The applications of this feature can be numerous, based on individual system requirements. One common application is to measure the static pressure difference across AHU filters and provide an alarm if the filters need to be cleaned or replaced. Similarly, for pumps, water flow rate and pressure can be monitored to initiate cleaning of pump strainers.

Another application of the alarm feature is to monitor the condenser approach temperature (difference between refrigerant temperature in the condenser and condenser water leaving temperature) of chillers and alert when it is necessary to clean the condenser tubes.

9.8 System Optimization and Control

9.8.1 Chiller controls

The basic control strategy for chillers involve the adjusting of refrigerant flow to the compressor in response to changes in cooling load to maintain the chilled water supply temperature (Fig. 9.7). In addition, chillers have additional control features to ensure their safe operation, such as flow sensors, to ensure sufficient chilled water and condenser water flow rates.

Chiller systems normally consist of chillers, pumps, and cooling towers, and advanced control systems are able to control and sequence operation of all equipment to ensure optimum system efficiency. Some of these sequencing strategies are described here.

Chiller sequencing for systems with only primary pumping. This strategy uses cooling load, chilled water temperature, and chiller motor loading to sequence the chillers (Fig. 9.8).

An additional chiller is turned on when the temperature of the chilled water leaving the chiller T_1 is greater than set point (chiller/s in operation cannot satisfy load) or if the cooling load is equal to the capacity of the operating chiller and chiller motors amps is equal or greater than full load amps (motors fully loaded).

An operating chiller will be switched off if the cooling load is less than the capacity of operating chillers, less the capacity of one chiller, and T_1 is not greater than set point.

This strategy for sequencing chillers takes into consideration the loading of chiller motors, which helps to ensure that chillers are fully loaded at off-design operating conditions, such as at lower condenser

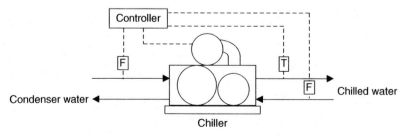

Figure 9.7 Basic control parameters for chillers.

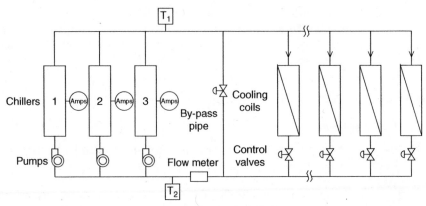

Figure 9.8 Chiller sequencing using cooling load, chilled water temperature, and motor loading.

water supply temperature and higher chilled water temperature, when chillers can provide higher than the rated capacity.

Strategy for chiller sequencing for systems with primary-secondary pumping. Primary-secondary systems normally need to be sequenced based on the chilled water flow in the decoupler pipe (Fig 9.9). If the flow is from the return side to the supply side, it indicates that the chilled water flow in the secondary loop is higher than that in the primary loop and, therefore, an additional primary pump (and a chiller) needs to be turned on. Similarly, if the temperature of chilled water leaving the chillers (T_1) is greater than set point, it indicates that the chillers in operation are unable to meet the cooling load and an additional chiller needs to be operated.

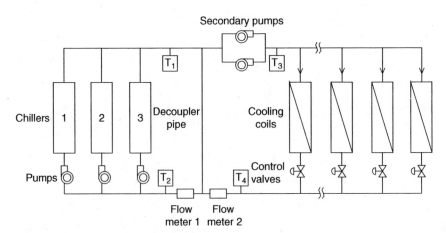

Figure 9.9 Chiller sequencing for primary-secondary systems.

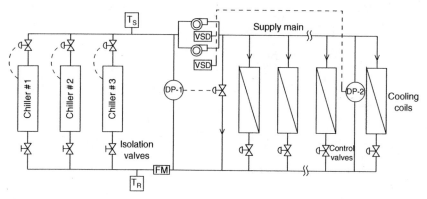

Figure 9.10 Variable primary flow systems.

If the flow in the decoupler is from supply side to return side and the flow is greater than 110 percent of the flow of one chiller, one chiller is switched off as the chilled water flow requirements can still be met by the remaining chillers.

Strategy for chiller sequencing for variable flow systems. The chillers are sequenced based on the cooling load computed using the flow meter (FM) and chilled water return and chilled water supply temperature sensors, T_R and T_S, respectively (Fig. 9.10). Motorized valves, interlocked to the chillers, are used to prevent chilled water circulation through chillers that are not in operation.

The VSDs of the chilled water pumps are controlled based on the differential pressure sensor DP-2 to maintain a set differential pressure across the furthest AHU coil. The motorized by-pass valve is controlled using differential pressure sensor DP-1 to maintain a minimum differential pressure across the chillers to ensure minimum flow. If the chilled water flow is too low, the differential pressure sensor DP-1 will sense that the pressure is below set-point and will open the bypass valve, enabling some water to bypass and circulate through the chillers. Due to this bypassing of chilled water, the differential pressure sensor DP-2 will sense a drop in pressure and will signal for the speed of the pumps to be increased. The system should be designed to ensure that the flow is maintained within minimum and maximum flow limits for chillers (usually 0.9–3.4 m/s). This can be achieved by using the flow meter installed on the primary circuit to fix the set point for the differential pressure sensor DP-1 and setting a minimum speed for the pumps on the VSDs.

9.8.2 Boiler controls

Control systems are required for boilers to ensure optimum fuel to air ratio for combustion and feedwater level control. A typically boiler

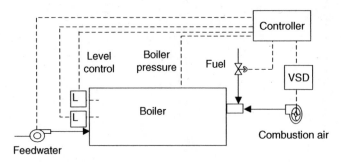

Figure 9.11 Typical controls for boilers.

control system is designed to maintain a set steam pressure irrespective of the steam usage (load), as shown in Fig. 9.11. This is achieved by varying the fuel flow to the boiler to vary the combustion rate in response to changes in the load. When the fuel flow rate is adjusted, the control system will automatically adjust the combustion airflow rate to ensure that the required air-to-fuel ratio is maintained. In more advance systems, the air-to-fuel ratio is also continuously adjusted based on the composition of flue gas to optimize the air-to-fuel ratio.

In boiler systems, the operation of feedwater pumps are also integrated with the boiler controls to maintain the water level in the boiler within the minimum and maximum levels.

9.8.3 Pump controls

Most pumps used in buildings are normally part of chiller or boiler systems and their operations are therefore controlled by the respective chiller or boiler control system. Other independent pumping systems can be controlled based on parameters such as tank level, temperature, and pressure or flow rate, as shown in Figs. 9.12 and 9.13.

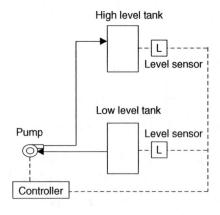

Figure 9.12 Typical level control system for pumps.

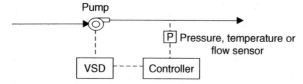

Figure 9.13 Typical pump control system.

9.8.4 AHU controls

In constant air volume (CAV) systems, the AHU control strategy is to modulate the chilled water flow through the coil in response to changes in cooling load. This is achieved by monitoring the return or space air temperature and varying the chilled water flow to maintain the return air temperature at a fixed value (Fig. 9.14). In such systems, the supply air temperature is effectively varied to maintain the space temperature at a fixed value.

In variable air volume (VAV) systems, the AHU control strategy is to maintain the supply air temperature and modulate the supply airflow rate in response to changes in cooling load. This is achieved by monitoring the static pressure in the supply air ducting and varying the speed of the fan (or inlet guide vanes in some AHUs) to maintain a fixed static pressure value (Fig. 9.15). The temperature of supply air is maintained constant by modulating the chilled water flow through the coil using a control valve.

9.8.5 FCU controls

In fan coil units, the space temperature is maintained at a set value by using a valve (normally the on/off type) to control chilled water flow

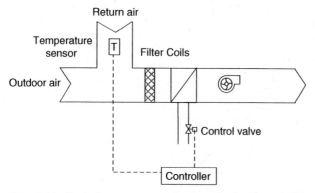

Figure 9.14 Control arrangement of a constant air volume AHU.

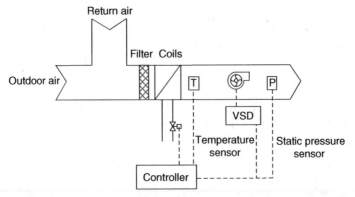

Figure. 9.15   Control arrangement of a variable air volume AHU.

through the cooling coil in response to changes in the cooling load (Fig. 9.16). FCU controls generally also have provision to adjust the fan speed.

9.8.6 Ventilation fan controls

The control of ventilation fans depends on the actual application. Generally, where fan capacity control is used, the requirement is to maintain a parameter such as CO_2 level, CO level, or temperature below a set maximum level in a space (Fig. 9.17). Sometimes, fan-capacity control is also used to maintain a set static pressure in a space in relation to another space or the atmosphere (Fig. 9.18).

9.8.7 Other optimization measures

Optimum start-stop.   The optimal start-stop algorithm available on most BAS can predict how long a building or space will take to cool down or heat up based on the variables that affect it, such as outdoor air

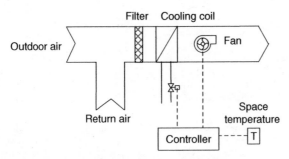

Figure 9.16   Control arrangement of a FCU.

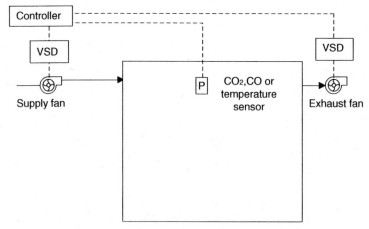

Figure 9.17 Ventilation control arrangement for maintaining a parameter in a space.

temperature, indoor space temperature, and building characteristics. The algorithm then predicts how long it will take to cool or heat the space and starts the chiller or boiler system and AHUs at the latest possible time to achieve the required space conditions before it is occupied.

Temperature reset for chillers. As explained in Chapter 2, chiller efficiency improves with increase in chilled water supply temperature. In general, it is estimated that improvement of 1 to 2 percent in chiller efficiency can be achieved by increasing the chilled water temperature by $0.6°C$ ($1°F$).

Chiller systems are generally designed for chilled water supply at $6.7°C$ ($44°F$) to meet the design peak cooling load. However, chillers

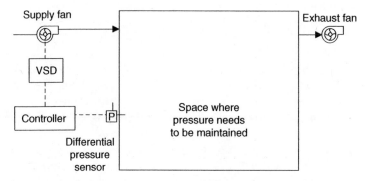

Figure 9.18 Ventilation control arrangement for maintaining static pressure in a space.

seldom have to operate under full load conditions. Therefore, most of the time it would not be necessary to provide chilled water at the design value and the chilled water temperature can be reset upwards.

One method of resetting the chilled water temperature is by monitoring the position of the control valves on the cooling coils using a BAS. The valve that is open most can be used to reset the temperature and the chilled water temperature can be reset upwards in steps until this control valve or any other reaches a preset value (eg. 90 percent open). Similarly, the chilled water temperature can be reset downwards if any control valve opens beyond the set maximum value.

Another method used for resetting the chilled water temperature utilizes the return chilled water temperature. When the return chilled water temperature reduces (e.g. from 12 to 10°C), the chilled water supply temperature is raised to bring the return temperature back to the design value (12°C). This is not a good strategy as the chilled water return temperature is the average for all AHUs and may not represent the loading of all AHUs. It could lead to situations where some areas experiencing full load conditions may not be able to provide sufficient cooling due to higher chilled water temperature.

The chilled water temperature can also be reset upwards based on the cooling load or outdoor temperature. Since reduction in outdoor temperature leads to lower cooling load, the outdoor temperature or a direct measurement of the load can be used to reset the chilled water temperature using a BAS.

Variable pumping systems.   Variable speed pumps are often used to vary the water flow rate to match load requirements. In such systems, the speed of the pumps are varied to maintain a set pressure differential in the system, which acts as the measure of building load.

Further energy savings can be achieved by varying the differential pressure set point used for controlling the VSD speed according to demand. In such a system, a BAS can be used to monitor the position of the control valves at the AHUs and reduce the differential pressure set point while ensuring that none of the valves are starved of water. A possible control strategy is shown in Fig. 9.19.

In this algorithm, the valve actuator position of all AHU chilled water modulating valves are monitored by the BAS and the valve that is open the most is determined. The opening of this valve is compared with limits set for adjusting the set point. For example, if the valve that is open the most is less than 70 percent open, it indicates that the other valves are open even less than 70 percent and, therefore, the pressure set point can be reduced further. Similarly, when load conditions change, if a valve is open more than 90 percent, the set point will be increased to prevent the cooling coil from being starved of chilled water.

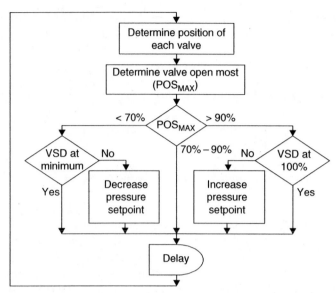

Figure 9.19 Control strategy for optimizing variable speed chilled water pumps.

Variable air distribution systems. Most commercial buildings use variable air volume (VAV) systems for air distribution where the amount of air provided to each space is varied based on load using thermostats and VAV boxes. Such VAV systems normally have VSDs fitted to their AHU fans to vary the amount of air supplied based on the load to save fan energy during part-load operation. This is achieved by modulating the fan speed to maintain a set static pressure in the distribution duct. The operations of VAV boxes and AHU fans are normally controlled by a BAS.

In VAV systems where the VAV box damper positions are monitored by a BAS, further energy savings can be achieved by continuously resetting the static pressure set point based on system operations. The static pressure set point can be continuously varied (as opposed to a fixed set point in normal systems) in response to the damper positions of the VAV boxes to ensure that no box is starved of air. This can help in minimizing AHU fan energy consumption by reducing the system static pressure while ensuring that the load requirements of all spaces are met.

A typical static pressure reset algorithm is similar to the one used for variable speed pumping systems (Fig. 9.19). The control algorithm is set to reduce the AHU fan speed up to the minimum set point while ensuring no VAV box is more than 90 percent open or is likely to be starved of air. The algorithm is able to achieve this by monitoring the damper positions of all the VAV boxes in a particular system and identifying the box with the damper open the most and, therefore, the most likely VAV box to be starved of air. If the box with the damper open most is more than 90 percent open, then the static pressure set point is raised. Similarly, if the box with the damper open most is open less than 70 percent (other boxes are open even less), the static pressure set point is reduced. This check is performed by the control system continuously to ensure that the system pressure is maintained at the minimum possible. It is estimated that such systems are able to achieve further 20 percent energy savings as compared to conventional fixed set point VAV systems.

Space temperature reset.   Air-conditioning systems maintain comfort conditions in occupied spaces by removing heat and moisture generated by occupants and equipment. The space temperature to maintain occupant comfort in a space is set based on parameters such as occupant activity, clothing worn by occupants, relative humidity of space, air circulation rate, and radiant heat gain into the space.

Therefore, for spaces where occupants are active, the space temperature is set lower to account for the extra heat that needs to be rejected by the bodies of the occupants. Similarly, for spaces with glazing (in perimeter zones of buildings), the space temperature is normally set lower to account for the radiant heat transfer between the surroundings and the occupants.

However, parameters such as the level of human activity and radiant heat gain can vary during different times of the day and during different periods of the year. Therefore, the space temperature set point can be continuously reset by the BAS to match such changes and help maintain comfort conditions. Since energy consumed by central chilled water cooling systems depend on the chilled water temperature, increasing space temperature set point can help to raise the chilled water temperature, which will result in better chiller efficiency.

Economizer cycle.   Another energy saving feature that can be incorporated into a BAS in some climates is the outside-air economizer. The basis of this strategy is to use outside air when it is below a certain temperature to cool the space rather than using a mixture of outside air and return air.

Normally, if the outdoor temperature is below the indoor temperature (return air temperature), the economizer cycle can be set to convert AHUs to use 100 percent outdoor air by adjusting the position of the outdoor air and return air dampers (Fig. 9.20). In humid climates, enthalpy based controls are preferred. Generally, parameters such as the outdoor temperature/enthalpy and indoor temperature are used for activating this energy saving strategy.

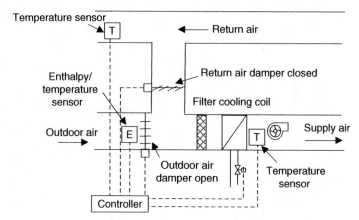

Figure 9.20 Arrangement of controls for a typical economizer.

9.9 Summary

Energy consumed by building systems such as air-conditioning, heating, and lighting systems can be reduced by ensuring that their operations are optimized by implementing various operating strategies. The chapter provided a summary of how such control and optimizing strategies can be implemented using the various functions and features of building automation systems.

Chapter

10

Building Envelope

10.1 Introduction

Buildings are designed to provide a comfortable internal environment for the occupants all year round despite variations in external weather conditions. This is achieved by using heating, ventilating, and air-conditioning (HVAC) systems.

The building envelope, which mainly consists of walls, roofs, windows, doors, and floors, allows heat to flow between the interior and exterior of a building and, hence, plays a key role in regulating the indoor environment. Therefore, the thermal characteristics of a building envelope has significant influence on heating, ventilating, and air-conditioning systems, and affect, both, equipment capacity and energy required for their operation.

Most measures to optimize the thermal performance of building envelopes need to be incorporated at the design stage of buildings or during a major upgrading exercise as they are not easy to implement and involve considerable expenditure. Although such improvement measures are relatively expensive, because they normally result in lower heating and cooling loads, which in turn results in downsizing of equipment and lower energy consumption, they are generally financially viable when considered on a life-cycle basis.

10.2 Envelope Heat Transfer

Buildings gain or lose heat by heat transfer or air leakage through the building envelope, as shown in Fig. 10.1. Heat transfer takes place by conduction, convection, and radiation, while air leakage occurs by infiltration and exfiltration.

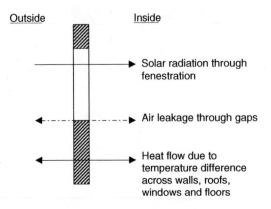

Figure 10.1   Heat gain/loss mechanisms for building envelopes.

Heat transfer due to conduction and convection occurs across walls, roofs, windows, and floors. It is dependent on the difference in temperature between the inner and outer surfaces of the envelope and can flow inwards or outwards. Heat transfer due to radiation takes place through fenestration and is mainly due to solar radiation, which flows inwards. Similarly, air leakage through building envelopes is dependent on the pressure difference between the interior and exterior of buildings and can be in either direction.

Conduction.   Heat transfer by conduction takes place when there is a temperature gradient across a solid object. The rate of heat transfer depends on the thermal conductivity of the material, its thickness, the temperature gradient, and the surface area available for heat transfer.

The rate of heat transfer by conduction can be expressed using Fourier's law as follows:

$$q_{cond} = -kA \frac{dT}{dx} \qquad (10.1)$$

where k = thermal conductivity of the material
    A = area (perpendicular to heat flow)
    dT = temperature gradient
    dx = thickness

In buildings, heat is transferred by conduction mainly through walls, roofs, and floors. Heat also travels by conduction through glazing of building envelopes.

Convection.   Heat transfer by convection takes place when a fluid comes into contact with a surface at a different temperature. Normally, in building walls, roofs, and windows, heat transfer by convection takes place at, both, the inner and outer surfaces.

Heat transfer by convection can be expressed using Newton's law of cooling, as follows:

$$q_{conv} = h_c A(T_s - T_f) \tag{10.2}$$

where h_c = surface heat transfer coefficient
  A = surface area
  T_s = surface temperature
  T_f = fluid temperature

The amount of convective heat transfer depends on the surface area, temperature difference between the surface and fluid, and the surface heat transfer coefficient. The surface heat transfer coefficient is dependent on the wind conditions for outdoor surfaces, while for indoor surfaces it is dependent on the airflow over the surface caused by HVAC systems.

Radiation. Heat transfer by radiation takes place due to electromagnetic waves, which travel at the speed of light. In buildings, radiant heat transfer is mainly due to the transmission of solar radiation through fenestration on the building envelope.

The amount of radiant heat transmission between two surfaces depends on the absolute surface temperatures of the bodies exchanging heat and the area of the body at the higher temperature. Radiant heat transfer can be expressed as follows:

$$q_{rad} = \sigma A_1 \varepsilon_1 (T_1^4 - T_2^4) \tag{10.3}$$

where σ = Stefan-Boltzmann constant
  A_1 = area of surface 1
  ε_1 = emissivity of surface 1
  T_1 = absolute temperature of surface 1
  T_2 = absolute temperature of surface 2

Radiant heat transfer can also be described by a simple expression using the radiant heat transfer coefficient (h_r), as follows:

$$q_{rad} = h_r A_1 (T_1 - T_2) \tag{10.4}$$

Since, typically, a major portion of the energy consumed by buildings is used for providing space cooling and heating, energy savings in buildings can be achieved by reducing the heat gain or loss by building envelope components. Some of the important aspects of building envelopes, which impact building energy consumption, are discussed in this chapter.

10.3 Walls and Roofs

Heat transmission through opaque building walls and roofs is illustrated in Fig. 10.2. Heat transfer by conduction takes place through the building fabric while convection heat transfer takes place at the inner and outer surfaces. The actual direction of heat conduction depends on the difference in temperature between the inner and outer surfaces and can vary from inwards during summer to outwards during winter.

Generally, walls are composites made up of a few layers of different materials, as shown in Fig. 10.3. Using Eq. (10.1) for each layer, the overall heat conduction can be expressed as:

$$q_{cond} = - \frac{A}{\left[\left(\frac{x}{k}\right)_{brick} + \left(\frac{x}{k}\right)_{insulation} + \left(\frac{x}{k}\right)_{plaster} \right]} (T_2 - T_1) \tag{10.5}$$

where k = thermal conductivity of each material
        A = area (perpendicular to heat flow)
        x = thickness of each layer
        $T_{1,2}$ = temperature of the outer and inner surfaces

Equation (10.5) can be simplified and expressed as:

$$q_{cond} = -U.A \, (T_2 - T_1) \tag{10.6}$$

where the overall heat transfer coefficient

$$U = \frac{1}{\left[\left(\frac{x}{k}\right)_{brick} + \left(\frac{x}{k}\right)_{insulation} + \left(\frac{x}{k}\right)_{plaster} \right]}$$

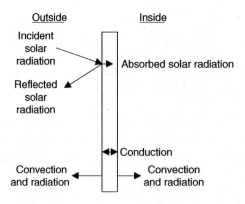

Outside   Inside

Incident solar radiation → Absorbed solar radiation

Reflected solar radiation

Conduction

Convection and radiation   Convection and radiation

Figure 10.2 Heat transfer through walls and roofs.

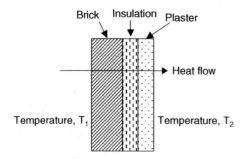

Brick Insulation Plaster

→ Heat flow

Temperature, T_1 Temperature, T_2

Figure 10.3 Heat transfer through a composite wall.

Therefore, heat transfer by conduction through walls and roofs is dependent on the thermal conductivity and thickness of the wall or roof (overall resistance or U value), the thermal gradient, and the surface area, as seen in Eq. (10.1).

As wall and roof surface areas are normally dependent on the building design while the thermal gradient is controlled by outdoor and indoor conditions, heat conduction through walls and roofs can generally only be reduced by increasing the overall thermal resistance. This can be achieved by using materials with lower thermal conductivity for walls and roofs or using insulation materials having low thermal conductivity, like fiberglass bats, rigid board insulation, and blown insulation.

Insulation is useful not only in cold climates, where it helps to reduce heat loss, it is also useful in climates that require air-conditioning as it can minimize heat gain. Increasing the insulation level of buildings to minimize heat losses or heat gain can be more easily achieved for new buildings by incorporating them into the design of buildings.

For roofs, since warm air moves upwards, the temperature difference between the interior and exterior is higher, leading to a higher rate of heat conduction. Therefore, roof and ceiling insulation is important to minimizing heat loss. In existing buildings, roof insulation can be improved relatively easily during reroofing exercises.

Since properties like texture and color of surfaces have an impact on radiant heat transfer, such properties of walls and roofs affect the heat loss or heat gain by buildings. Generally, smooth, light colored surfaces reflect more solar energy than dark, rough surfaces. Hence, for buildings that are air-conditioned, smooth and light colored surfaces are preferred while for heated buildings, dark and rough surfaces are desirable. For buildings that require heating during winter and cooling during summer, the preferred color for the roof will depend on whether the building has greater heating or cooling needs during the major part of the year.

10.4 Windows

Glazed or fenestrated windows make a significant contribution to the heat exchange between the building's conditioned space and outdoors. In tropical climates, where buildings are air-conditioned, glazing can transmit as much heat as ten to twenty times the equivalent area of wall surface. Since heat gain due to solar radiation has a bigger influence on the cooling load of air-conditioned buildings than heat conduction and long-wave radiation, in hot climates, most energy conservation measures for fenestration are aimed at reducing heat gain by solar radiation. Similarly, in heated buildings, as solar heat gain through fenestration actually helps to reduce the heating load, energy saving measures for fenestration in such buildings are mainly aimed at minimizing conduction losses.

Heat flow through a typical window is illustrated in Fig. 10.4. As shown in the figure, heat transfer takes place due to conduction and radiation. The amount of radiant heat transfer depends on the amount of incident direct solar radiation (the location of glazing), type of glass, and surface area, while the conduction heat transfer depends on the thermal conductivity, thickness, and surface area of the glazing.

·Some of the important properties that provide a measure of window performance are U-value, solar heat gain coefficient, shading coefficient, and visible transmittance.

U-value. The U-value is a measure of heat flow through a window due to temperature difference between the exterior and interior of buildings. It is similar to the U-value in Eq.(10.6) for composite walls and can account for the different layers in double and triple glazed windows. The units of U-value are $W/m^2.K$ and the heat flow through windows is proportional to the U-value. Therefore, the heat flow through a window with a U-value of 1.0 will be double that for a window with a U-value of 0.5.

The U-value of a window depends on the type of glazing, its thickness, coatings used, type of gas (air or inert gas), and thickness of gap in

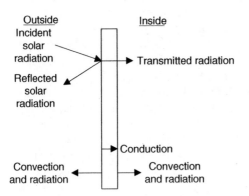

Figure 10.4 Heat flow through a typical window.

multipaned windows. Typical U-values range from 0.9–1.1 for single glazed to 0.4–0.5 for double glazed and 0.2–0.3 for triple glazed windows.

Solar heat gain coefficient (SHGC).   The solar heat gain coefficient (SHGC) is a measure of how much of the solar radiation incident on a window is transmitted through it. SHGC can range from 0 to 1, where SHGC = 0 means none of the radiation incident on the window is transmitted through as heat while SHGC = 1 means all the incident radiation is transmitted through the window.

Since lower SHGC results in lower solar heat gain through the glazing from the exterior to the interior, windows with low SHGC are preferred for air-conditioned buildings while windows with high SHGC are preferred for buildings using passive solar heating.

Typical SHGC values range from 0.8 for single glazed to 0.6 for double glazed and 0.4 for triple glazed windows.

Shading coefficient (SC).   The shading coefficient (SC) is another term used to measure the solar transmittance of glass. It is the ratio of heat transmittance of a particular glass to the heat transmittance through a 1/8 inch clear glass.

Visible transmittance.   The visible transmittance (T_{vis}) is a measure of how much of the visible light incident on a window is transmitted through it. A typical clear, single-pane window has a T_{vis} of 0.90, which means that it will allow 90 percent of the visible light incident on it to pass through.

Shading devices.   Various shading devices can be used to minimize the transmission of direct solar radiation through windows. Shading devices are generally classified as internal or external shading devices. External shading devices, such as louvers and overhangs, installed outside the building envelope help to reduce transmission of direct radiation into the building by intercepting direct radiation before it reaches the glazing (Fig. 10.5). However, they also cut off a portion of natural light, which may lead to higher lighting loads for perimeter areas of buildings.

Interior shading devices, such as curtains and blinds, can also be used but are less effective at reducing solar gains than the external shading devices since the solar radiation first enters the conditioned space and is then reflected back through the glazing before being absorbed by the conditioned space (Fig. 10.6).

Special coatings.   Since the properties of glazing can also be altered by tinting or by applying various coatings or films, special coatings and solar control films can be used to reduce solar heat gain through windows.

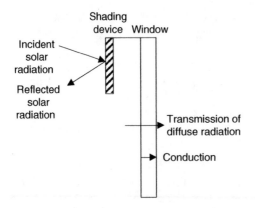

Figure 10.5 External shading device.

These include tinted glass and spectrally selective coatings that can be used to transmit visible light while reflecting part of the solar radiation.

Glass tints are generally achieved by either adding colors during production or by applying colored films after production. Although tints are normally able to reduce solar gains by absorbing part of the solar radiation, they normally reduce the transmission of natural light. Another disadvantage is that most of the absorbed heat energy is subsequently transferred into the building through convection and radiation.

Coatings usually have microscopically thin metallic or ceramic coatings to reduce the transmission of solar radiation while allowing part of the visible light to be transmitted. They are ideal for energy conservation as heat gain can be minimized while maximizing the use of natural light. They can be added during production or applied in the form of a film for existing windows.

The percentage of solar energy rejected and the percentage of visible light allowed to be transmitted depends on the actual type of coating or film used. Typical infrared rejection and visible light transmission properties

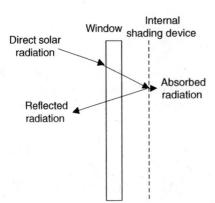

Figure 10.6 Internal shading device.

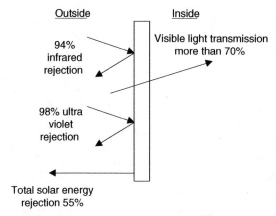

Figure 10.7 Performance of a typical solar film.

of a high performance film is shown in Fig. 10.7. Since the solar spectrum consists of visible light, ultraviolet and infrared, part of the energy in solar radiation is in the form of visible light. Therefore, when such a special film is able to reject even more than 90 percent of infrared and ultraviolet, the total effective solar energy rejection is about 55 percent.

Multipaned windows. While minimizing radiant heat gain through glazing is the prime concern in climates that require building air-conditioning, heat loss through glazing is important in buildings that require heating to maintain occupant comfort. In such applications, the insulation property of the glazing can be improved by using double or triple glazing, which consists of two or three layers of glass sandwiching layers of air (or gas) between them to make use of the low thermal conductivity of air (gas) to reduce heat transmission, as shown in Fig. 10.8 for a double glazed window.

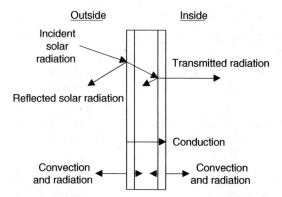

Figure 10.8 Heat flow through a typical double glazed window.

10.5 Air Leakage

Air leakage through building envelopes can be a cause of energy wastage in buildings. Leakage can occur due to infiltration, which is the leaking of untreated air into the building and exfiltration, which is the leaking out of treated air.

Excessive air infiltration occurs due to negative pressure in buildings. This can be caused by unbalanced supply and exhaust airflows, where the exhaust airflow is much higher than the supply airflow rate. Typical reasons for such unbalanced airflows can be the design or sizing of the ventilation system and reduction of supply air to reduce building cooling or heating loads.

Exfiltration is the opposite of infiltration and can also be caused by unbalanced supply and exhaust flows where supply airflow is higher than exhaust airflow.

Infiltration and exfiltration can also occur in buildings due to wind effects. Wind pressure is positive on the surface of the building facing the wind while it is negative on the opposite face, leading to air passing through the building. Similarly, stack effect in buildings take place where cold air from the outside flows in at the lower levels of the building to take the place of warm air rising within the building (which in turn leaks out at the higher levels).

Some common means of reducing such air leakage through building structures include weather-stripping, use of automatic doors, vestibules, and air curtains.

Weather-stripping.   Weather-stripping involves the sealing of gaps around exterior doors and windows to prevent air leakage. Different types of weather-stripping are available for different applications. The common types of weather-stripping used are compression type for normal doors and windows and sliding type for sealing sliding surfaces.

Automatic doors.   Automatic doors can be used in applications where the doors need to be frequently opened. Such doors can be completely automatic, that is, they are opened and closed automatically using sensors, which can detect the movement of people or doors that have hydraulic or spring loaded mechanisms to simply close the doors when opened.

Vestibules.   Vestibules are intermediate compartments that act as air locks between the exterior and interior of buildings to minimize air leakage at doorways. They usually have two sets of automatic or revolving doors so that the interior of the building is not directly exposed to the exterior.

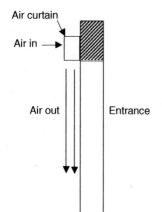

Air curtain

Air in →

Air out   Entrance

Figure 10.9   Operation of a typical air curtain.

Air curtains.   Air curtains can be installed over open doorways to prevent air leakage and are commonly used in retail stores where the doors are kept completely open (Fig 10.9). They consist of fans that take air from the front and blow it vertically downwards to create an invisible air barrier between the interior and exterior of the building.

10.6   Overall Thermal Transfer Value (OTTV)

Building codes in various jurisdictions generally specify minimum performance criteria for building envelope components to address energy conservation needs. These codes normally specify minimum design parameters for thermal resistance and solar radiation transmission for walls, roofs, floors, and fenestration.

An alternative approach is to specify a maximum allowable heat gain for the entire envelope of buildings and provide the designers the freedom to choose individual components and materials. The *overall thermal transfer value* (OTTV) used in some countries follows this approach.

The OTTV concept takes into consideration the basic elements of heat gain through the envelope of a building, which are:

- Heat conduction through opaque walls
- Heat conduction through fenestration
- Solar radiation through fenestration
- Heat conduction through roofs

The OTTV for walls is defined as:

$$OTTV = (TD_{eq} \times (1\text{-}WWR) \times U_w) + (\Delta T \times WWR \times U_f) + (SF \times CF \times WWR \times SC)$$

where WWR = window to wall ratio
U_w = thermal resistance of wall (W/m^2·K)
U_f = thermal resistance of fenestration (W/m^2·K)
TD_{eq} = equivalent temperature difference (K)
ΔT = temperature difference
SF = solar factor (W/m^2)
CF = correction factor for solar heat gain through fenestration
SC = shading coefficient for fenestration

A similar OTTV computation for roofs will include heat conduction through the opaque parts of the roof and skylights and solar radiation through skylights.

The various weather dependent factors used in the OTTV calculation, such as SF, TD_{eq}, and ΔT are derived from local weather data. The maximum OTTV value set for some countries is 45 W/m^2. This maximum value is periodically reviewed and revised downwards based on technological developments, which lead to improvements in the thermal properties of materials used in building envelopes.

10.7   Estimation of Building Energy Performance

At the building design stage or when retrofitting buildings, an accurate evaluation of the thermal performance of building envelope components is necessary to assist the selection of cost-effective options. Often, the available options need to be evaluated not just for their thermal performance but also for their effect on the overall energy performance of the building.

Since building envelopes are exposed to outdoor weather conditions, which are rarely steady, and because components such as walls and roofs also exhibit a thermal storage effect, numerical techniques such as finite element, finite difference, and time series, which can account for the transient nature of the heat flow, need to be used to predict the heat flow through building components.

The *thermal response factor method,* or *transfer function method,* is one such method where heat flow through walls is expressed algebraically in terms of the surface temperature history and thermal characteristics of the wall material using time series techniques. The *cooling load temperature difference* (CLTD) method uses factors developed by the methodology and equations of the transfer function method for direct one-step calculation of heat gain through sunlit walls, roofs, and glass exposures.

One of the most commonly used building energy analysis codes is DOE-2, which can be used to predict the energy use and cost for different types of buildings. DOE-2 uses a description of the building layout, constructions, usage, and conditioning systems like HVAC, lighting, and utility rates provided by the user, along with weather data, to perform an hourly simulation of the building.

Various easy to use software such as PowerDOE, which operate using DOE-2 as a platform, are also available to help perform quick and accurate predictions of the thermal and energy performance of buildings. Such computer modeling programs are very useful tools for building designers as they are able to optimize designs by simulating the thermal performance of individual components or the overall energy performance of buildings.

Results of a parametric study carried out using DOE-2 to evaluate the impact of various key parameters on the total energy usage and cooling energy usage of a typical building in the tropics is shown in Table 10.1. The table lists the variation in total building energy consumption and cooling energy consumption for a 10 percent increase in various parameters within a set minimum and maximum range. For each, two values are listed where the first value refers to an increase in the minimum range value while the other refers to a change in the maximum range value.

For example, a 10 percent increase in the COP from the maximum range value of 5.5 will result in a reduction in the total energy consumption and cooling energy consumption by 2.5 and 7 percent, respectively (negative sign indicate reductions).

However, one of the main drawbacks of such simulation tools is that they are normally not able to take into consideration the impact of rain on the building envelope. Since rain is absorbed by porous building materials, such as walls and roofs, which is later evaporated during drying, it leads to a reduction in the actual heat flow due to latent heat absorption, resulting in less accurate prediction of thermal and energy performance of buildings. The impact of this effect can be significant for buildings in climates that experience rainfall during most parts of the year.

In such situations, if accurate data is required, correction factors can be used to adjust the output data from standard simulation tools to

TABLE 10.1 Impact of Changes in Various Parameters on Total and Cooling Energy Consumption of a Building

Parameter	Range considered	Change in total energy consumption	Change in cooling energy consumption
COP of cooling plant	3.5 to 5.5	−3.5% to −2.5%	−7.9% to −7%
Lighting power (W/m^2)	10.8 to 20.4	+3.6% to +5.1%	+1% to +2%
Window-to-wall ratio	0.3 to 0.6	+0.82% to +0.81%	+1.6% to +1.5%
Shading coefficient	0.35 to 0.69	+1.3% to +2.3%	+2.4% to +4%
Cooling set point (°C)	22.3 to 27.8	−1.5% to −7%	−6% to −12%

account for the effect of rain. One study on this subject has shown that on rainy days, the actual heat gain through porous walls can reduce by 10 to 20 percent, while the annual reduction in heat gain for a building could be as high as 5 percent.

10.8 Summary

The building envelope, which consists of walls, roof, windows, doors, and floor, acts as the barrier between the outdoor environment and the conditioned indoor environment. This helps to minimize the load on HVAC systems used in buildings to condition the indoor environment. In general, HVAC systems account for a significant portion of the total energy consumed in buildings. Therefore, the overall energy consumption of buildings can be reduced by improving the thermal design of building envelopes, and various strategies to achieve this were described in the chapter.

Bibliography

ASHRAE, *Handbook of Refrigeration*, American Society of Heating, Refrigerating and Air-conditioning Engineers, Inc., Atlanta, GA, 2002.

ASHRAE, *Handbook of HVAC Applications*, American Society of Heating, Refrigerating and Air-conditioning Engineers, Inc., Atlanta, GA, 2003.

ASHRAE, *Handbook of Fundamentals*, American Society of Heating, Refrigerating and Air-conditioning Engineers, Inc., Atlanta, GA, 2005.

ASHRAE Standard 90.1, Energy Standard for Buildings Except Low-Rise Residential Buildings, American Society of Heating, Refrigerating and Air-conditioning Engineers, Inc., Atlanta, GA., 2001.

ASHRAE Standard 62.1, Ventilation for Acceptable Indoor Air Quality, American Society of Heating, Refrigerating and Air-conditioning Engineers, Inc., Atlanta, GA., 2004.

Berbari, G.J., "Fresh Air Treatment in Hot and Humid Climates," *ASHRAE Journal*, 64–70, 1998.

Chou, S.K. and Y.K. Lee, "A Simplified Overall Thermal Transfer Value Equation for Building Envelopes Energy," *The International Journal*, 13(8), 657–670, 1988.

Chou, S.K. and W.L. Chang, "Effects of Multi-Parameter Changes on Energy Use of Large Buildings," *International Journal of Energy Research*, Vol. 17, 885–903, 1993.

Chou, S.K., N.E. Wijeysundera, and S.E.G Jayamaha, "Determining the Heat Flow Through Building Walls Under Simulated Actual Weather Patterns," *International Journal of Energy Research*, Vol. 19, 243–251, 1995.

Chou, S.K. and W.L Chang, "A Generalized Methodology for Determining the Total Heat Gain Through Envelopes," *International Journal of Energy Research*, Vol. 20, 887–901, 1996.

Cleaver Brooks, *Boilerspec*, WebCD, Version 1.0.

Desiccant Rotors International/Bry-Air product catalogue.

DOE-2 Engineers Manual, US Department of Commerce, National Technical Information Center,1980.

Eppelheimer, D.M., "Variable Flow–The Quest for System Energy Efficiency," *ASHRAE Transactions*, 96 (12), 673–678.

Jayamaha, S.E.G., "Thermal Response of Building Materials and Components Under Hot and Humid Conditions," MEng thesis, National University of Singapore, 1993.

Jayamaha, S.E.G., N.E. Wijeysundera, and S.K. Chou, "Measurement of the External Heat Transfer Coefficient for Walls Under Outdoor Weather Conditions," Proceedings of the Asia Pacific Conference on the Built Environment, Vol. 1, 498–509, 1995.

Jayamaha, S.E.G., "Heat and Moisture Transfer Through Building Envelope Components Subjected to Outdoor Weather Conditions Including Rain," PhD thesis, National University of Singapore, 1996.

Krarti, M., *Energy Audit of Building Systems—An Engineering Approach*, CRC Press, Florida, USA ,2000.

Mashuri, W. and L.K. Norford, "Integrating VAV Zone Requirements with Supply Fan Operation," *ASHRAE Journal*, 43–46, April 1993.

Moult, R and J. Tran, "Getting an Additional 20% Energy Savings from VAV Systems," ASHRAE Conference—Built Environment Trends and Challenges, Singapore, June 1995.

National Lighting Product Information Program (NLPIP), *Specifier Report* 1(2), 1992.

National Lighting Product Information Program (NLPIP), *Specifier Report* 6(2), 1998.

Philips Lamps and Gear Catalogue 2005/2006.

Spirax Sarco Learning Centre (www.spiraxsarco.com/learn/).

Schanin, D.J., Plug Loads—The Fastest Growing Consumer of Commercial Power, Bayview Technology Group, Inc.

SS CP 13: Code of Practice for Mechanical Ventilation and Air-conditioning in Buildings, PSB, 1999.

SS CP 24: Code of Practice for Energy Efficiency Standard for Building Services and Equipment, PSB, 1999.

SS CP 38: Code of Practice for Artificial Lighting in Buildings, PSB, 1999.

Tillack, L and J.B. Rishel, "Proper Control of HVAC Variable Speed Pumps," *ASHRAE Journal*, 41–47, November 1998.

Wang, K. S., *Handbook of Air-conditioning and Refrigeration*, McGraw Hill, Singapore, 2001.

Wijeysundera, N.E., S.K. Chou, and S.E.G. Jayamaha, "Heat flow Through Walls Under Transient Rain Conditions," *Journal of Thermal Insulation and Building Envelopes*, Vol. 17, 118–141, 1993.

Wulfinghoff, D.R., *Energy Efficiency Manual*, Energy Institute Press, Maryland, USA, 1999.

Solutions

Chapter 2

2.1.

    (i) Chilled water reset, which is to increase the chilled water supply temperature set point for chillers. Increasing the chilled water set point helps to reduce the compressor lift (pressure difference across compressor), thereby helping to improve chiller efficiency.

    (ii) Condenser water reset, which is to reduce the condenser water supply temperature to the chillers. Reducing the condenser water supply temperature helps to reduce the condenser water return temperature, which also results in the reduction of compressor lift, which, in turn, leads to better chiller efficiency.

    (iii) Matching chiller capacity to cooling load. The operating efficiency of chillers is generally dependent on the loading (ratio of cooling load to chiller capacity), and higher loading results in better efficiency.

    (iv) For water-cooled chillers, reducing the condenser approach temperature (temperature difference between refrigerant in the condenser and water leaving the condenser) also helps improve chiller efficiency by reducing the compressor lift. This can be achieved by condenser tube cleaning and a good water treatment program.

2.2. VSDs can be used to vary the speed of chilled water pumps to better match their capacity to the cooling load. The pump capacity can be varied to maintain a set pressure in the chilled water distribution system.

2.3. Efficiency of the chiller is 0.6 kW/RT, when the chilled water supply temperature is 7°C.

    If the chilled water supply temperature is increased to 8°C, the chiller efficiency will improve by 3 percent to 0.6 kW/RT × 0.97 = 0.58 kW/RT. If the chiller operates at 500 RT, 10 hrs a day,

$$\text{Saving} = (0.6 - 0.58) \times 500 \times 10 = 100 \text{ kWh/day}$$

2.4.

$$\begin{aligned} \text{Energy cost} = {} & \text{cooling load} \times \text{hours of operation} \\ & \times \text{efficiency (kW/RT)} \times \text{electricity tariff} \end{aligned}$$

For the 0.5 kW/RT chiller,

$$\text{the energy cost for Year 1} = 500 \text{ RT} \times 10 \text{ h/day} \times 250 \text{ days/year} \\ \times 0.5 \text{ kW/RT} \times \$0.10/\text{kWh} = \$62,500$$

$$\text{The energy cost for Year 2 with 2 percent escalation} \\ \text{of tariff is } \$63,750 \ (\$62,500 \times 1.02)$$

The annual life-cycle cost comparison for the three chillers can be computed as follows:

| | Annual cost for chillers | | |
	0.5 kW/RT	0.55 kW/RT	0.65 kW/RT
Year 0	$300,000	$275,000	$250,000
Year 1	$62,500	$68,750	$81,250
Year 2	$63,750	$70,125	$82,875
Year 3	$65,025	$71,528	$84,533
Year 4	$66,326	$72,958	$86,223
Year 5	$67,652	$74,417	$87,948
Year 6	$69,005	$75,906	$89,707
Year 7	$70,385	$77,424	$91,501
Year 8	$71,793	$78,972	$93,331
Year 9	$73,229	$80,552	$95,197
Year 10	$74,693	$82,163	$97,101
Total	$984,358	$1,027,793	$1,139,665

(All costs are energy costs except Year 0 capital cost)

2.5.

$$\text{Daily saving} = (1.75 - 0.65) \times 20 \text{ RT} \times 8 \text{ h}$$

$$= 176 \text{ kWh/day}$$

2.6.

(i) Daily kWh savings can be computed as follows:

Time	Hours	Cooling load (RT)	Chiller efficiency (kW/RT)	Chiller kWh
12 a.m. to 6 a.m.	6	300	0.57	1026
6 a.m. to 10 a.m.	4	400	0.56	896
10 a.m. to 2 p.m.	4	450	0.555	999
2 p.m. to 8 p.m.	6	400	0.56	896
8 p.m. to 12 p.m.	4	350	0.565	791
			Daily total	4608

Present chiller consumption = 6000 kWh/day

$$\text{Saving} = 6000 - 4608 = 1392 \text{ kWh/day}$$

$$\text{Daily savings} = 1392 \text{ kWh}$$

(ii) Tariff = $0.10/kWh

$$\text{Operating days/year} = 300$$

$$\text{Annual savings} = 1392 \times 0.1 \times 300 \times \$41{,}760$$

$$\text{Payback} = \text{cost/savings} = 300{,}000/41{,}760 = 7.2 \text{ years}$$

Chapter 3

3.1. If the oxygen concentration in the flue gas is reduced from 8 to 5 percent, in Fig. 3.9, the combustion efficiency will increase by about 2 percent.

3.2. The temperature difference between the flue gas and room temperature is 142°C (approximately 290°F).

Using the stack loss Table A.2 (for no.2 oil) in Appendix A, for CO_2 concentration of 5.5 percent in the flue gas, the stack loss is estimated to be 20.9 percent.

From Table 3.1, the radiative and convective losses are about 0.5 to 0.7 percent.

Therefore, the overall efficiency = $100 - (20.9 + 0.7) = 78.4\%$

3.3. Savings can be estimated as follows:

Boiler loading A	Operating hours a day B	Fan motor power with damper (kW) C	Fan motor power with VSD (kW) $D = A^3 \times 35$	Power saving (kW) $E = C - D$	Energy savings (kWh) $B \times E$
80%	4	22	18	4	16
60%	12	18	8	10	180
40%	8	16	2.2	13.8	17.6
			Total daily savings		213.6

3.4.

Make-up water 27°C and x unit

Condensate 95°C and $(1-x)$ units ⟶ [ ] ⟶ Feedwater 80°C and 1 unit

Heat balance,

$$27\,x + (1-x)\,95 = 80$$

$$x = 0.22 \text{ (22 percent makeup water)}$$

Amount of condensate is recovered $= (1 - x) = 0.78$
(78 percent condensate is recovered).

3.5. From Fig. 3.17, at 300 kPa, the approximate steam leak rate for a 3 mm hole is 10 kg/hr. The amount of fuel that can be saved is 6000 litres a year (based on 24 hours × 7 days a week × 50 weeks).

For three holes, the total leakage will be 18,000 litres a year. Annual fuel cost savings that can be achieved if these leaks are eliminated = 18,000 × \$0.5 = \$9000.

Chapter 4

4.1.

$$Q_1 = 200 \text{ L/s}$$
$$P_1 = 45 \text{ kW}$$
$$N_1 = 1200 \text{ rpm}$$
$$N_2 = 1100 \text{ rpm}$$

$$Q_2 = Q_1 \times (N_2/N_1) = 200 \times (1100/1200) = 183 \text{ L/s}$$
$$P_2 = P_1 \times (N_2/N_1)^3 = 45 \times (1100/1200)^3 = 35 \text{ kW}$$

4.2. From pump affinity laws, the pump speed can be reduced to give the design flow as follows:

$$\text{New pump speed} = 1200 \times (20/35) = 686 \text{ rpm}$$

$$\text{New power consumption} = 25 \times (20/35)3 = 4.7 \text{ kW}$$

$$\text{Reduction in pump power consumption} = 25 - 4.7 = 20.3 \text{ kW}$$

4.3.

	Option A	Option B
Pressure drop across condenser	80 kN/m^2	40 kN/m^2

The theoretical pump power consumption to overcome the resistance across the condenser can be calculated as follows:

Pump kW = [Flow in m^3/s $\times$ pressure difference in N/m^2]/1000

Option 1, pump kW = $[0.150 \times 80 \times 10^3]/1000 = 12$ kW

Option 2, pump kW = $[0.150 \times 40 \times 10^3]/1000 = 6$ kW

Pump kW savings = $(12 - 6) = 6$ kW.

4.4.

$$P = \frac{(Q \times \Delta P)}{(1000 \times \eta_p \times \eta_m)}$$

If the motor efficiency η_m is taken as 1.0,

$$\text{Saving in pump power} = \frac{(Q \times \Delta P)}{1000} \left(\frac{1}{\eta_{78\,\%}} - \frac{1}{\eta_{90\%}} \right)$$

$$= \frac{(0.04 \times 120 \times 10^3)}{1000} \left(\frac{1}{0.65} - \frac{1}{0.85} \right)$$

$$= 1.7 \text{ kW}$$

Chapter 5

5.1.

Possible causes	Remedies
Defective water spray mechanism	Repair spray system
Defective/damaged fill material	Repair/replace fill material
Inadequate airflow	Tighten/replace drive belts

5.2. The wet-bulb and dry-bulb temperatures of air entering the cooling tower are much higher than the ambient wet-bulb and dry-bulb temperatures; possibly due to recirculation of air discharged from the cooling tower back to the air intake.

5.3. At full load, the fan consumes 15 kW.

At 60 percent speed (60 percent capacity), from the cube law,

$$\text{Fan kW} = 15 \times (0.6)^3 = 3.2 \text{ kW}$$

$$\text{Savings} = (15 - 3.2) = 11.8 \text{ kW}$$

5.4.

Motor power of one fan = 15 kW (at full speed)

If the fan operates at 50 percent speed, fan motor power

$$= 15 \times (0.5)^3 = 1.9 \text{ kW}$$

Total fan power when both fans operate at 50 percent speed

$$= 1.9 \times 2 = 3.8 \text{ kW}$$

Fan power savings = $(15 - 3.8) = 11.2 \text{ kW}$

5.5.

Estimated free cooling capacity of each 500-RT cooling tower

$$= 25 \text{ RT } (500 \times 0.05)$$

Estimated free cooling capacity of each 200-RT cooling tower

$$= 10 \text{ RT } (200 \times 0.05)$$

Total free cooling capacity = $(25 \times 3) + (10 \times 2) = 95 \text{ RT}$

Chapter 6

6.1.

$Q_1 = 15 \text{ m}^3/\text{s}$

$Q_2 = 10 \text{ m}^3/\text{s}$

$P_1 = 30 \text{ kW}$

$N_1 = 1400 \text{ rpm}$

$$N_2 = N_1 \times (Q_2/Q_1) = 1400 \times (10/15) = 933 \text{ rpm}$$

$$P_2 = P_1 \times (N_2/N_1)^3 = 30 \times (933/1400)^3 = 8.9 \text{ kW}$$

6.2.

Fan impeller power (kW)

$$= \frac{\text{flow rate (m}^3/\text{s)} \times \text{pressure developed (N/m}^2\text{or Pa)}}{1000 \times \text{efficiency}}$$

Taking the efficiency as 1.0,

$$\text{Fan power (kW)} = \frac{20 \times (80 - 30)}{1000 \times 1.0}$$

$$= 1 \text{ kW}$$

6.3. From the psychrometric data in Appendix B, the enthalpy of outdoor air is 90 kJ/kg (at 35°C dry-bulb at 60 percent RH), and the enthalpy of return air is 47.5 kJ/kg (at 23°C and 55 percent RH).

$$\begin{aligned}\text{Reduction in cooling load} &= \text{airflow rate} \times \text{density of air} \\ &\quad \times \text{difference in enthalpy}\end{aligned}$$

$$= 1.5 \text{ m}^3\text{/s} \times 1.2 \text{ kg/m}^3 \times (90 - 47.5) \text{ kJ/kg}$$

$$= 76.5 \text{ kW (approximately 22 RT)}$$

$$\begin{aligned}\text{Savings in chiller power consumption} \\ = \text{cooling load reduction} \times \text{system efficiency}\end{aligned}$$

$$= 22 \text{ RT} \times 0.8 \text{ kW/RT}$$

$$= 17.6 \text{ kW}$$

$$\text{Annual energy consumption} = 17.6 \text{ kW} \times 12 \text{ h/day} \times 250 \text{ days/year}$$

$$= 52{,}800 \text{ kWh/ear}$$

6.4.

Using Eq. (6.6),

$$T_{SA} = T_{OA} - (T_{OA} - T_{RA}) \times \eta_T$$

$$= 33 - (33 - 23) \times 0.77$$

$$= 25.3°C$$

Sensible cooling done,

$$Q_{sensible} = 1.232 \times v \times (T_{OA} - T_{SA})$$

$$= 1.232 \times 3 \times (33 - 25.3)$$

$$= 28.5 \text{ kW}$$

$$= 8 \text{ RT}$$

Savings in chiller power consumption

$$= \text{cooling load reduction} \times \text{system efficiency}$$
$$= 8\ \text{RT} \times 0.83\ \text{kW/RT}$$
$$= 6.6\ \text{kW}$$

Annual energy consumption $= 6.6\ \text{kW} \times 10\ \text{h/day} \times 260\ \text{days/year}$
$$= 17{,}160\ \text{kWh/year}$$

Chapter 7

7.1.

(i) Delamping (to reduce lux level to about 500)

(ii) Changing from T8 to T5 lamps (need to replace fixture or use T5 with adaptors)

(iii) Using power saving device and dimming to achieve about 500 lux

7.2. Use of timers

a. Office lighting

b. Outdoor lighting

Use of occupancy sensors

a. Toilets

b. Meeting rooms

Use of light sensors to dim lighting

a. Perimeter office lighting

b. Outdoor/car park lighting

7.3.

Energy savings per lamp $= (60 - 11) = 49\ \text{W}$

Energy savings per day $= (49 \times 1000 \times 12)/1000\ \text{kWh}$
$$= 588\ \text{kWh/day}$$

Annual energy savings $= 588 \times 5 \times 52 \times \$0.10 = \$15{,}288$

Payback period based purely on energy savings

$$= \$(12 \times 1000)/\$15{,}288 = 0.78 = 9.4\ \text{months}$$

Payback period taking into account longer lamp life:

Annual cost of incandescent lamps = $1 × (12 × 5 × 52)/1000 5× $3.12

Annual cost of CFL = $12 × (12 × 5 × 52)/8000 = $4.68

Total additional cost = $(4.68 − 3.12) × 1000

$$= \$1,560$$

Payback period = 1560/15,288 = 0.1 = 1.2 months

7.4.

Energy savings per lamp = (40 − 28) = 12 W

Energy savings per day = (12 × 100 × 24)/1000 kWh

$$= 28.8 \text{ kWh/day}$$

Annual savings = 28.8 × 365 × $0.1 = $1051.20

Total cost = $40 × 100 = $4000

Payback period = 4000/1051.20 = 3.8 years

7.5.

Energy savings per day = (84–64) × 24)/1000 kWh = 0.48 kWh/day

Annual energy savings = 0.48 × 365 × $0.10 = $17.52

Cost of ballast = $20

Payback period = 20/17.52 = 13.7 months

Chapter 8

8.1.

$$\text{For the motor, Power}_{\text{IN}} = \frac{\text{Power}_{\text{OUT}}}{\eta}$$

$$\text{Energy saved in kWh} = \text{Motor kW} \times \text{Operating hour} \times \left[\frac{1}{\eta_S} - \frac{1}{\eta_P} \right]$$

where η_S = efficiency of standard motor
η_P = efficiency of premium motor

Annual electrical saving

$$= 55 \times 24 \times 250 \left[\frac{1}{0.91} - \frac{1}{0.95} \right] \text{kWh/year}$$

$$= 15{,}269 \text{ kWh/year.}$$

8.2.

Energy saved in kWh

$$= \text{Transformer load (kVA)} \times \text{pF} \times \text{Operating hour} \times \left[\frac{1}{\eta_{98.5}} - \frac{1}{\eta_{99.5}} \right]$$

where pF = power factor
η = efficiency of transformers

Annual electrical saving

$$= 1{,}500 \times 0.9 \times 12 \times 250 \left[\frac{1}{0.985} - \frac{1}{0.995} \right] \text{kWh/year}$$

$$= 41{,}323 \text{ kWh/year.}$$

8.3.

a. Chilled water distribution pumps

b. Hot water distribution pumps

c. Cooling tower fans

d. Air handling unit fans

e. Supply and exhaust ventilation fans

8.4.

a. Replace hydraulic elevators with traction elevators

b. Replace old motor drives with modern VVVF drives

c. Install drives that can feedback regenerated energy back into the distribution system

d. Install controls to optimize multiple elevator systems

e. Install controls to switch off systems when elevator is not in service

f. Reduce counterweight and elevator capacity (if elevator capacity is more than required)

8.5.

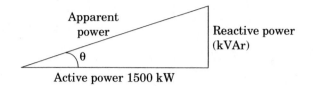

Reactive power at 0.75 power factor

$$\cos \theta = 0.75$$

$$\theta = 41.41°$$

Reactive power = $\tan (41.41°) \times 1500 = 1323$ kVAr

Reactive power at 0.9 power factor

$$\theta = \cos^{-1}(0.9) = 25.8°$$

Reactive power = $\tan (25.8°) \times 1500 = 725$ kVAr

Therefore, size of capacitor required

$$= 1323 - 725 = 598 \text{ kVAr (capacitive)}.$$

Reference Data for Boiler Operations

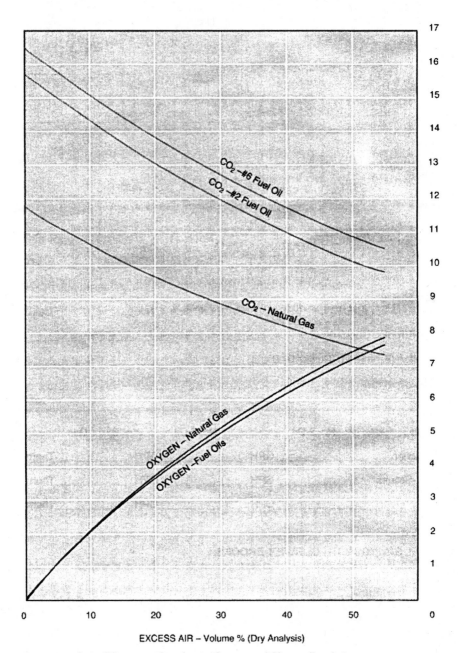

Figure A.1 O_2 to CO_2 conversion chart. (*Courtesy of Cleaver Brooks.*)

TABLE A.1 Stack Loss Estimation Table for Natural Gas

STACK LOSS - % - NATURAL GAS

DIFFERENCE BETWEEN FLUE GAS AND ROOM TEMPERATURES IN DEGREES FAHRENHEIT

% CO_2	200	220	240	260	280	300	320	340	360	380	400	420	440	460	480	500	520	540	560	580	600	620	640	660	680	700	750	800	850	900	950	1000
3.0	23.1	24.4	25.9	27.2	28.6	30.0	31.3	32.8	34.1	35.8	36.9	38.2	39.8	41.0	42.2	43.8	45.0	46.3	47.8	49.0	50.0											
3.5	21.2	22.5	23.8	24.9	26.1	27.2	28.4	29.6	30.9	32.0	33.2	34.4	35.8	36.8	37.9	39.2	40.3	41.6	42.8	43.8	45.0	46.2	47.7	48.3	49.8							
4.0	19.9	20.9	22.0	23.1	24.1	25.1	26.2	27.2	28.3	29.4	30.4	31.8	32.5	33.8	34.8	35.8	36.8	37.8	38.8	39.9	40.9	42.1	43.0	44.1	45.2	46.2						
4.5	18.9	19.9	20.9	21.8	22.7	23.6	24.5	25.5	26.4	27.3	28.3	29.2	30.2	31.2	32.2	33.0	34.0	34.9	35.9	36.8	37.8	38.6	39.8	40.4	41.5	42.6	44.8	47.2				
5.0	18.0	18.9	19.8	20.6	21.4	22.2	23.1	24.0	24.9	25.8	26.8	27.5	28.3	29.1	30.1	30.9	31.8	32.5	33.6	34.3	35.7	36.2	36.9	37.8	38.8	39.7	41.8	43.8	46.0	48.2		
5.5	17.4	18.1	18.9	19.8	20.5	21.2	22.1	22.9	23.8	24.5	25.2	26.2	26.9	27.8	28.5	29.2	30.0	30.8	31.8	32.3	33.2	34.1	34.9	35.8	36.3	37.3	39.2	41.0	43.0	45.3	47.2	49.0
6.0	16.8	17.4	18.2	18.9	19.6	20.4	21.1	21.8	22.7	23.3	24.1	24.9	25.5	26.2	27.0	27.8	28.4	29.2	30.0	30.8	31.5	32.2	32.9	33.8	34.3	35.2	36.8	38.8	40.4	42.5	44.3	46.2
6.5	16.3	16.9	17.6	18.4	19.0	19.8	20.4	21.1	21.8	22.4	23.2	23.8	24.5	25.2	25.9	26.5	27.2	27.9	28.7	29.2	30.0	30.9	31.4	32.1	32.8	33.5	34.6	36.8	38.4	40.3	42.0	43.8
7.0	15.8	16.5	17.1	17.8	18.4	19.1	19.8	20.4	21.0	21.8	22.3	22.9	23.6	24.2	24.9	25.5	26.2	26.8	27.4	28.0	28.8	29.4	30.0	30.8	31.2	32.0	33.8	35.3	36.8	38.3	40.0	41.8
7.5	15.5	16.1	16.7	17.2	17.9	18.5	19.1	19.8	20.3	20.9	21.5	22.2	22.8	23.3	24.0	24.6	25.2	25.8	26.4	26.9	27.7	28.2	28.8	29.4	30.1	30.8	32.2	33.8	35.2	36.8	38.3	39.9
8.0	15.2	15.7	16.3	16.9	17.4	18.0	18.6	19.2	19.8	20.3	20.9	21.5	22.1	22.8	23.2	23.8	24.4	25.0	25.5	26.0	26.7	27.2	27.8	28.4	29.0	29.5	31.0	32.4	33.8	35.4	36.8	38.2
8.5	14.9	15.4	15.9	16.5	17.1	17.6	18.2	18.7	19.3	19.8	20.4	20.9	21.4	22.0	22.5	23.1	23.7	24.2	24.8	25.3	25.8	26.4	26.9	27.4	28.1	28.6	29.9	31.3	32.8	34.2	35.4	36.8
9.0	14.6	15.2	15.7	16.2	16.6	17.2	17.8	18.3	18.8	19.3	19.9	20.4	20.9	21.4	21.9	22.5	23.0	23.5	24.1	24.5	25.2	25.8	26.2	26.7	27.2	27.8	29.0	30.3	31.8	33.0	34.3	35.7
9.5	14.4	14.9	15.4	15.9	16.4	16.9	17.4	17.9	18.4	18.9	19.5	19.9	20.5	20.9	21.4	21.9	22.4	22.9	23.4	23.8	24.4	24.9	25.4	25.9	26.4	26.9	28.2	29.4	30.8	32.0	33.3	34.5
10	14.2	14.6	15.2	15.5	16.1	16.6	17.1	17.5	18.1	18.5	19.0	19.5	20.0	20.4	20.8	21.4	21.8	22.4	22.8	23.3	23.8	24.2	24.8	25.2	25.8	26.2	27.4	28.6	29.8	31.2	32.2	33.4
11		14.4	14.7	15.2	15.6	16.1	16.5	16.9	17.4	17.8	18.4	18.8	19.3	19.6	20.2	20.5	20.9	21.4	21.9	22.3	22.8	23.2	23.7	24.2	24.6	25.0	26.2	27.2	28.3	29.5	30.8	31.8
12			14.4	14.8	15.2	15.6	16.1	16.5	16.9	17.3	17.8	18.2	18.6	19.0	19.4	19.8	20.2	20.6	21.1	21.4	21.9	22.3	22.8	23.2	23.6	24.0	25.1	26.1	27.2	28.3	29.2	30.3

Source: Courtesy of Cleaver Brooks.

TABLE A.2   Stack Loss Estimation Table for No. 2 Oil

STACK LOSS - % - NO. 2 OIL

DIFFERENCE BETWEEN FLUE GAS AND ROOM TEMPERATURES IN DEGREES FAHRENHEIT

% CO_2	200	220	240	260	280	300	320	340	360	380	400	420	440	460	480	500	520	540	560	580	600	620	640	660	680	700	750	800	850	900	950	1000
3.0	24.1	25.8	27.7	29.3	31.3	33.9	34.8	36.4	38.2	40.0	42.9	44.8	45.5	47.0	49.0	50.8	52.4	54.3	56.0	57.9	59.6	61.5	63.5	65.0	66.8	68.8						
3.5	21.7	23.1	24.8	26.2	27.8	29.2	31.7	32.5	33.9	35.3	36.9	38.5	40.0	41.7	43.1	44.8	46.1	47.8	49.4	50.9	52.2	53.9	55.7	57.0	58.3	60.0	63.8	67.8				
4.0	19.9	21.2	22.5	24.9	25.2	26.5	27.9	29.2	31.7	32.0	33.3	35.8	36.0	37.3	38.7	40.0	41.4	42.9	44.1	45.5	46.9	48.1	49.8	50.9	52.1	53.8	57.0	60.2	63.9	67.1		
4.5	18.4	19.7	20.8	22.0	23.2	24.4	25.6	26.9	28.0	29.3	30.4	31.8	32.9	34.2	35.6	36.7	37.8	39.0	40.1	41.2	42.5	43.8	45.0	46.3	47.4	48.8	51.8	54.6	57.8	60.9	63.9	66.9
5.0	17.2	18.5	19.5	20.7	21.7	22.7	23.8	24.9	26.0	27.1	28.2	29.4	30.3	31.5	32.7	33.8	34.9	35.9	36.8	38.0	39.2	40.1	41.7	42.4	43.7	44.7	47.4	50.1	52.9	55.8	58.3	61.2
5.5	16.3	17.4	18.4	19.4	20.4	21.3	22.3	23.3	24.3	25.4	26.3	27.3	28.4	29.4	30.6	31.4	32.4	33.6	34.5	35.3	36.4	37.4	38.4	39.6	40.3	41.7	44.0	46.5	49.0	51.8	54.1	56.5
6.0	15.6	16.5	17.4	18.3	19.3	20.4	21.2	22.0	23.0	23.9	24.9	25.8	26.8	27.7	28.6	29.5	30.4	31.4	32.3	33.1	34.2	35.0	36.0	36.9	37.9	38.9	41.0	43.5	45.8	48.0	50.3	52.8
6.5	14.9	15.7	16.7	17.5	18.4	19.3	20.1	20.9	21.8	22.7	23.6	24.5	25.3	26.1	27.0	27.8	28.8	29.6	30.6	31.3	32.3	33.0	34.1	34.8	35.7	36.5	38.7	40.8	42.9	45.1	47.5	49.7
7.0	14.4	15.3	16.0	16.8	17.8	18.4	19.3	20.1	20.9	21.8	22.4	23.2	24.1	24.9	25.7	26.5	27.3	28.1	28.9	29.8	30.5	31.4	32.3	33.0	33.8	34.6	36.5	38.6	40.5	42.7	44.7	46.6
7.5	13.9	14.6	15.4	16.2	16.9	17.7	18.5	19.2	20.1	20.7	21.3	22.2	23.0	23.8	24.5	25.2	26.0	26.8	27.5	28.2	29.0	29.8	30.6	31.3	32.2	32.9	34.8	36.5	38.5	40.3	42.3	44.2
8.0	13.5	14.3	14.9	15.7	16.3	17.1	17.7	18.5	19.3	20.0	20.7	21.4	22.1	22.8	23.5	24.2	25.0	25.7	26.3	27.0	27.8	28.5	29.2	30.0	30.8	31.5	33.2	35.0	36.8	38.5	40.2	42.1
8.5	13.2	13.8	14.5	15.2	15.8	16.5	17.3	17.8	18.6	19.3	20.0	20.6	21.3	21.9	22.6	23.3	23.9	24.6	25.3	25.9	26.7	27.3	28.0	28.8	29.4	30.1	31.8	33.5	35.2	36.9	38.7	40.2
9.0	12.8	13.4	14.1	14.7	15.4	16.0	16.7	17.3	17.9	18.6	19.3	20.0	20.6	21.2	21.8	22.4	23.1	23.8	24.4	25.0	25.7	26.3	27.0	27.7	28.3	28.9	30.5	32.1	33.8	35.3	37.0	38.5
9.5	12.5	13.2	13.7	14.3	14.9	15.7	16.3	16.8	17.4	18.1	18.6	19.3	19.9	20.5	21.1	21.7	22.4	22.9	23.5	24.1	24.8	25.4	26.0	26.7	27.2	27.9	29.4	31.0	32.5	34.0	35.5	37.2
10	12.3	12.8	13.4	14.0	14.6	15.2	15.7	16.3	16.9	17.5	18.1	18.7	19.3	20.0	20.5	21.0	21.6	22.2	22.8	23.4	24.0	24.6	25.1	25.8	26.3	27.0	28.3	29.9	31.4	32.9	34.4	35.7
11	11.8	12.4	12.8	13.4	13.9	14.5	15.0	15.5	16.2	16.7	17.2	17.8	18.3	18.7	19.4	20.0	20.5	20.9	21.5	22.0	22.6	23.1	23.7	24.2	24.8	25.3	26.7	28.0	29.4	31.8	32.1	33.5
12	11.4	11.8	12.5	12.9	13.4	13.9	14.4	14.9	15.4	15.9	16.4	16.9	17.4	17.9	18.4	18.9	19.5	20.0	20.5	20.9	21.4	21.9	22.4	22.9	23.5	24.0	25.2	26.5	27.8	29.0	30.2	31.7
13	11.2	11.6	12.1	12.5	12.9	13.4	13.9	14.3	14.7	15.3	15.8	16.3	16.7	17.2	17.7	18.1	18.6	19.1	19.6	20.1	20.5	21.1	21.3	21.8	22.3	22.8	24.0	25.2	26.3	27.5	28.8	30.0
14		11.3	11.8	12.2	12.6	13.0	13.4	13.8	14.3	14.8	15.3	15.6	16.2	16.5	16.9	17.4	17.8	18.3	18.7	19.2	19.7	20.2	20.6	21.0	21.4	21.8	22.9	24.1	25.2	26.2	27.4	28.6
15			11.4	11.7	12.4	12.6	13.1	13.5	13.8	14.3	14.8	15.3	15.6	15.9	16.4	16.7	17.3	17.7	18.1	18.4	18.9	19.4	19.8	20.3	20.6	21.0	22.0	23.1	24.2	25.2	26.2	27.3

Source: Courtesy of Cleaver Brooks.

Table A.3 Stack Loss Estimation Table for No. 6 Oil

STACK LOSS - % - NO. 6 OIL

DIFFERENCE BETWEEN FLUE GAS AND ROOM TEMPERATURES IN DEGREES FAHRENHEIT

% CO_2	200	220	240	260	280	300	320	340	360	380	400	420	440	460	480	500	520	540	560	580	600	620	640	660	680	700	750	800	850	900	950	1000
3.0	24.5	26.5	28.5	30.2	32.2	34.5	36.5	38.2	40.4	42.2	44.4	46.4	48.2	50.0	52.3	54.3	56.3	58.2	60.3	62.0	64.1	66.2	68.1	70.1								
3.5	21.8	23.4	25.2	26.8	28.6	30.4	32.1	33.8	35.5	37.4	39.0	40.6	42.2	44.0	45.6	47.5	49.2	51.0	52.2	53.9	55.7	57.0	58.3	60.0			63.8	67.8				
4.0	19.8	21.2	22.8	24.2	25.7	27.3	28.8	30.2	31.6	32.5	34.8	36.3	37.8	39.4	40.8	42.2	43.8	45.1	46.9	48.2	49.8	51.2	52.9	54.2	55.7	57.0	61.1	65.0	68.9			
4.5	18.2	19.4	20.8	22.2	23.5	24.8	26.2	27.4	28.8	30.4	31.5	33.0	34.2	35.4	37.0	38.1	39.4	41.0	42.2	43.5	45.0	46.3	47.9	49.0	50.1	51.9	55.0	58.2	61.8	65.1		
5.0	16.8	18.0	19.2	20.4	21.7	22.8	24.2	25.3	26.6	27.8	29.0	30.3	31.4	32.6	33.8	35.3	36.2	37.5	38.8	39.8	41.0	42.3	43.8	44.9	46.1	47.5	50.1	53.6	56.3	59.8	62.3	65.8
5.5	15.8	16.8	18.0	19.2	20.3	21.3	22.5	23.5	24.6	25.8	26.9	28.0	29.2	30.2	31.4	32.5	33.5	34.7	35.8	37.0	37.9	39.2	40.1	41.3	42.3	43.8	46.1	49.1	52.0	54.7	57.8	60.1
6.0	14.8	15.8	16.9	18.0	19.0	20.0	21.1	22.0	23.1	24.2	25.2	26.3	27.3	28.2	29.3	30.3	31.3	32.3	33.5	34.3	35.3	36.5	37.5	38.3	39.7	40.5	43.0	45.8	48.2	50.9	53.5	56.0
6.5	14.3	15.2	16.1	17.1	18.0	18.9	19.9	20.8	21.8	22.8	23.7	24.6	25.5	26.5	27.5	28.5	29.4	30.4	31.4	32.3	33.4	34.3	35.1	36.1	37.1	38.0	40.2	42.8	45.1	47.6	49.9	52.1
7.0	13.5	14.4	15.3	16.2	17.1	17.9	18.8	19.7	20.6	21.5	22.4	23.3	24.2	25.0	25.8	26.8	27.7	28.6	29.4	30.2	31.2	32.2	33.0	33.9	34.9	35.8	37.9	40.1	42.1	44.4	46.8	49.0
7.5	13.0	13.8	14.6	15.5	16.3	17.3	18.0	18.8	19.7	20.5	21.4	22.2	22.9	23.7	24.6	25.4	26.3	27.2	27.9	28.8	29.6	30.5	31.2	32.1	33.0	33.9	35.9	37.9	40.0	42.0	44.1	46.1
8.0	12.5	13.3	14.1	14.8	15.7	16.4	17.3	18.0	18.8	19.6	20.4	21.2	21.9	22.7	23.5	24.2	25.0	25.8	26.6	27.4	28.2	29.0	29.9	30.6	31.5	32.1	34.1	36.0	38.0	40.0	41.9	43.9
8.5	12.2	12.8	13.6	14.4	15.1	15.7	16.6	17.3	18.0	18.7	19.6	20.3	21.0	21.6	22.5	23.3	23.9	24.7	25.5	26.2	26.8	27.6	28.2	29.1	29.9	30.8	32.6	34.2	36.2	38.0	39.9	41.8
9.0	11.7	12.4	13.2	13.8	14.6	15.3	15.9	16.6	17.4	18.1	18.8	19.5	20.2	20.8	21.6	22.3	22.9	23.7	24.4	25.0	25.7	26.5	27.1	27.9	28.7	29.4	31.1	32.9	34.6	36.3	38.0	39.9
9.5	11.4	12.1	12.7	13.4	14.1	14.7	15.4	16.0	16.7	17.5	18.1	18.7	19.4	20.0	20.7	21.4	22.1	22.8	23.5	24.0	24.7	25.4	26.1	26.8	27.5	28.1	29.8	31.2	33.2	34.9	36.4	38.1
10	11.2	11.7	12.3	13.0	13.7	14.4	14.8	15.5	16.2	16.8	17.5	18.2	18.7	19.4	20.0	20.6	21.3	21.9	22.6	23.2	23.8	24.5	25.1	25.8	26.4	27.0	28.7	30.1	31.8	33.5	35.0	36.7
11	10.6	11.3	11.8	12.4	12.9	13.5	14.2	14.7	15.3	15.8	16.5	16.9	17.6	18.2	18.8	19.4	20.0	20.6	21.2	21.8	22.3	22.9	23.5	24.1	24.8	25.2	26.8	28.1	29.8	31.2	32.5	34.1
12	10.2	10.7	11.3	11.7	12.3	12.8	13.4	13.8	14.5	15.1	15.6	16.2	16.7	17.2	17.8	18.3	18.8	19.4	19.9	20.4	21.0	21.6	22.1	22.7	23.1	23.8	25.0	26.4	28.1	29.1	30.5	31.9
13		10.3	10.8	11.3	11.8	12.3	12.8	13.3	13.8	14.4	14.8	15.4	15.8	16.3	16.8	17.3	17.9	18.4	18.9	19.3	19.8	20.4	20.9	21.4	21.9	22.4	23.8	24.9	26.2	27.5	29.1	30.0
14		9.8	10.4	10.8	11.4	11.8	12.3	12.8	13.3	13.7	14.3	14.7	15.2	15.6	16.2	16.6	17.1	17.5	18.0	18.5	18.8	19.4	19.9	20.4	20.9	21.2	22.5	23.7	24.9	26.1	27.2	28.5
15			10.2	10.6	11.0	11.4	11.8	12.4	12.7	13.3	13.7	14.2	14.6	15.0	15.4	15.8	16.4	16.8	17.3	17.7	18.2	18.6	19.0	19.5	19.9	20.3	21.5	22.6	23.8	24.9	25.9	27.1
16			10.3	10.7	11.1	11.5	11.8	12.3	12.8	13.3	13.7	14.0	14.4	14.6	15.0	15.3	15.7	16.2	16.6	16.9	17.4	17.9	18.2	18.8	19.1	19.5	20.6	21.6	22.7	23.8	24.8	25.9

Source: Courtesy of Cleaver Brooks.

Appendix

B

Enthalpy of Moist Air

The following table lists the approximate enthalpy of moist air in kJ/kg at various Dry-bulb temperatures and relative humidity values.

Temperature (°C)	40% RH	45% RH	50% RH	55% RH	60% RH	65% RH	70% RH
15	25.7	27.0	28.4	29.7	31.0	32.4	33.8
16	27.4	28.8	30.3	31.7	33.1	34.6	36.0
17	29.1	30.7	32.2	33.8	35.3	36.8	38.4
18	31.0	32.6	34.2	35.9	37.5	39.2	40.8
19	32.8	34.5	36.3	38.0	39.8	41.6	43.3
20	34.7	36.6	38.4	40.3	42.2	44.0	45.9
21	36.7	38.6	40.6	42.6	44.6	46.6	48.6
22	38.7	40.8	42.9	45.0	47.1	49.3	51.4
23	40.7	43.0	45.2	47.5	49.8	52.0	54.3
24	42.9	45.3	47.7	50.1	52.5	54.9	57.3
25	45.1	47.6	50.2	52.7	55.3	57.9	60.5
26	47.3	50.0	52.7	55.5	58.2	60.9	63.7
27	49.7	52.5	55.4	58.3	61.2	64.1	67.1
28	52.1	55.1	58.2	61.3	64.4	67.5	70.6
29	54.5	57.8	61.1	64.3	67.6	70.9	74.2
30	57.1	60.5	64.0	67.5	71.0	74.5	78.0
31	59.7	63.4	67.1	70.8	74.5	78.2	82.0
32	62.5	66.4	70.3	74.2	78.2	82.1	86.1
33	65.3	69.4	73.6	77.7	81.9	86.2	90.4
34	68.2	72.6	77.0	81.4	85.9	90.4	94.9
35	71.2	75.9	80.6	85.3	90.0	94.7	99.5
36	74.4	79.3	84.3	89.2	94.3	99.3	104.4
37	77.6	82.8	88.1	93.4	98.7	104.1	109.5
38	81.0	86.5	92.1	97.7	103.3	109.0	114.7
39	84.4	90.3	96.2	102.2	108.2	114.2	120.3
40	88.1	94.3	100.5	106.8	113.2	119.6	126.0

Index

Entries denoted by an italic *t* or *f* indicate figures or tables respectively.